PLANT SCIENCE RESEARCH AND PRACTICES

# "*IN SITU*" BIOMONITORING OF A POLLUTED ENVIRONMENT BY WILD PLANT AND CROP PLANT SPECIES

# Plant Science Research and Practices

Additional books and e-books in this series can be found on Nova's website under the Series tab.

PLANT SCIENCE RESEARCH AND PRACTICES

# "*IN SITU*" BIOMONITORING OF A POLLUTED ENVIRONMENT BY WILD PLANT AND CROP PLANT SPECIES

KAROL MIČIETA

AND

GUSTÁV MURÍN

nova
science publishers
New York

**Library of Congress Cataloging-in-Publication Data**

ISBN: 978-1-53614-276-1

*Published by Nova Science Publishers, Inc. † New York*

# CONTENTS

# PREFACE

The original method of testing for pollution in the environment by means of wild and crop plant species is described, with practical examples explained in detail within this book.

This original method was developed by authors and proved for industrial complexes like nickel plant dumps, a smelting plant and an aluminium factory as well as city environments experiencing a heavy traffic impact.

For a demonstration of general usefulness, a collaborative study with Kuwait University was accomplished to show the impact of the first Gulf War to the environment.

The study of radioactive contamination in the vicinity of the Jaslovské Bohunice nuclear power plant was finished in cooperation with the University of Wien, Austria.

Results of phytotoxicity and mutagenicity of spots contaminated by low-radioactive bombs used by NATO in Serbia are reported here for a first time.

Promising results from the retrospective biomonitoring of polluted environments by means of pollen grains for herbal samples are also presented, detailing the situation at a chosen locality up to one hundred years ago.

A list of wild and crop plant species as bioindicators is presented for practical use.

This original method is extended by special chapter describing authors' another original "Parallel plant test of phytotoxicity and mutagenicity",

Practical examples of remediation of heavily polluted sites are included too.

In conclusion, the authors are underlying the remarkable ability of Nature to recover from all anthropogenic interventions thanks to the mechanisms of auto-regulation, adaptation and resistance.

As this book is based on the thirty years of the authors' experiences, it covers the period between the years of 1966-2018, with 399 references from this field of research.

# ACKNOWLEDGMENTS

This monograph was reviewed by:

***Dr. Agáta Fargašová***
Katedra Environmentálnej Ekológie PríF UK,
Mlynská Dolina, Bratislava, Slovakia

***Dr. S.A. Geras'kin,***
Laboratory of Radioecology
Russian Academy of Agricultural Sciences, Moscow, Russia

***Prof. Kateřina Malachová***
Katedra biologie a ekologie, Přírodovědecká fakulta OU,
Chittussiho 10, 701 03 Slezská Ostrava, Czech Republic

***Dr. Ewa Chudzińska***
Institute of Experimental Botany
Adam Mickiewicz University, Poznań, Poland

It has to be underlined that this method was originally an idea of Professor Augustín Murín who was supervising also our first steps in confirming its practical use.

At a first place we would like to thank to all of our collaborators who were in their parts following our research and participating with their capability at particular topics:

Abraham, D. (Kuwait), Afzal, M. (Kuwait), Attia, T. (Kuwait), Banásová, V., Bikar, T. (Serbia), Boulos, L. (Kuwait), Briggs, D. (Cambridge, UK), Brutovská, E., Bukovská, E., Dannford, N. (UK), Dienes, K., Dušička, J., Fargašová, A., Gregušková, E., Halašová, E., Holub, Z. (Czech Rep.), Hutár, V., Chrenová, J., Kerekeš, E., Knasmüller (Austria), S., Kovařík, P. (Czech Republic), Kubalová, S., Kunová-Staples, K., Madilová, J., Malallah, G., Mišík, M. (Austria), Polák, R., Pisarčíková, H., Pogányová, A., Ostrovský, I., Rendeková, A., Sámelová, A., Slávik, O., Solenská, M., Stano J., Svetková, K., Ščevková, J., Šípková, Ž., Terenová, A., Vršková, K., Zamečníková, M. and Záhradníková, E.. Zuzana Randáková made a great contribution to this book by formatting a manuscript and by her revision of the tables.

Authors thanks also to prof. Robert Murray Davis (US), Dr. Alice Botai (Canada) and Leslie Horwitz (US), for their revision of the text.

This study was supported by Vedecká grantová agentúra VEGA [1/0380/13] and VEGA [1/0885/16] and is also partially the result of the project implementation of Comenius University in Bratislava Science Park supported by the Research and Development Operational Programme funded by the ERDF Grant number: ITMS 26240220086 and Kuwait University, Research Grant No. SO 054.

*Chapter 1*

# INTRODUCTION

Environmental pollution is a global issue touching every society on Earth. The interest in environmental issues and particularly in environmental pollution is obviously based on the idea of possible threats to human health. The distance from the human body to plants seems to be a long one. However, both share the basic structure of DNA. Moreover, higher plants have already shown their ability to fulfill all requisite conditions for the bioindication of genotoxicity of environmental chemicals (cf. De Serres 1978, De Serres1992, Grant and Zura 1982, Grant et al. 1981, Ma 1990, Grant 1994, Burton 1989, Geras'kin et al. 2011a). There is no doubt that they could hardly compete with routine methods of testing mutagens and/or carcinogens by means of bacterial and mammalian material, although plant tests are cheaper and approximately 60-70% of hazardous chemicals have the same effects on plant tests as on bacterial and mammalian tests (Constantin and Owens 1982). For the evaluation of environmental chemicals, competent monitoring and an indication of their impact on the environment are required (Bromberg 1990). The best tools for it are bioindicators because the characterization of ecotoxicology factors through chemical analysis is limited (Holsen et al. 1991).

Therefore, at the end of the 20$^{th}$ century an international effort was focused on the standardization of certificated tests of genotoxic effects of xenobiotics. However, the research in this field was considerably more promising during that time.

## 1.1. History of the Research in This Field

The first methods for utilizing higher plants for tests of genotoxicity were nearly all "*in vivo*" tests (cf. Grant and Zura 1982, Vig 1982, Ma 1990, Van't Hof and Schairer 1982, Plewa 1982, Posthumus 1983, Murín 1984, Sandhu et al. 1994 a, b). A highlight of these attempts was a collaborative study of WHO/IPCS where most of the abovementioned laboratories, including our own, were involved. Outcomes of this project were four short-term bioassays – *Allium*, *Vicia*, *Arabidopsis* and *Tradescantia*. They were defined for practical use in a rapid screening of chemicals in laboratories with basic equipment and standardized cultivating conditions (Kanaya et al. 1994, Sandhu et al. 1991a).

However, the basic advantage, as we consider it – "*in situ*" bioindication of polluted environment via wild plant and crop plant species– was not utilized. Although a theoretical base for it was established in reports of De Serres (1978), Mulcahy (1981), Grant and Zura (1982), Bromberg (1990) and Bulgakov et al. (2003), several authors reported "*in situ*" studies with formerly "*in vivo*" standardized plant species, such as tobacco Bell $W_3$ (e.g., Posthumus and Tonnejick 1982), *Allium* (e.g., Grant 1982, Fiskesjö 1985, Liman et al. 2011), *Tradescantia* (e.g., Ma 1982, Sandhu et al. 1991b, Ichikawa 1992, Cebulska-Wasilewska 1992, Knasmüller et al. 1992, Mišík et al. 2011), *Vicia faba* (e.g., Mičieta 1987), *Nicotiana* plants (e.g., Chakraborty and Mukherjee 2009), maize (e.g., Mišík 2007) or their combination (e.g., Grant et al. 1992). Interesting results were obtained from a radioactive contamination in Chernobyl on *Arabidopsis* (e.g., Abramov and Schevchenko 1992), *Allium* (e.g., Geras'kin et al. 2011c) and *barley* seeds (e.g., Geras'kin et al. 2017).

Nonetheless, only a few reports were focused on the cytogenetic monitoring of natural (not artificially introduced for testing) vegetation "*in situ*" affected by pollutants such as herbicides (Tomkins and Grant 1976), polymers (Sharma and Raju 1983, Sandhu et al. 1991a), heavy metals (Udosen et al. 2006, Fargašová 2012, Pavlova 2016) or photochemical smog (Grant and Zura 1982, Cela Renzoni et al. 1990). For this type of contamination, a whole spectra of bioindicators were used (see Hraška et al. 2000, Geras'kin et al. 2017, Lourenço et al. 2016) including different types of vines (Aroutiounian et al. 2004a). Unique is evaluation of the impact of pollution on plant-*Lepidoptera* relationships (Mulder et al. 2005). The use of plants as models for tests of environmental pollution was reported in a monograph by Fargašová (2009). Paradiž and Lovka (1999) stressed importance of pollen grain bioassay for environmental pollution screening.

This was an inspiration for our research and the starting point for our objective of finding a new tool for practical testing of an environmental pollution. After more than thirty years of our own studies based on an Austro-Slovakian, a Serbian, a Kuwaiti and several Slovak projects, we would like to summarize our results to show the ability of wild and crop plant species to serve as suitable bioindicators of a polluted environment.

## 1.2. Preview of Our Partial Reports

Our previous experiences with "*in vivo*" tests of phytotoxic and mutagenic effects of polluted waters (Murín et al. 1980) and WHO/IPCS collaborative study for the standardization of certificated plant tests on the genotoxic effects of xenobiotics (Kanaya et al. 1994) focused our interest towards "*in situ*" bioindication. Initial reports by Murín A. (1987, 1995) stated the basic criteria for selection of plant bioindicators from the regional flora for monitoring of environmental pollution. First 100 species of Slovak flora were examined from which 40 wild species of native flora (Murín 1987), 21 species of cultivated crop plant species (Mičieta 1989) and 15 species of Kuwaiti flora (Murín and Malallah 1994, Malallah et al. 1997) were selected as bioindicators. For more see 2.3.

The problems arising from the pollution of the environment have led us to develop effective and in the same time low-cost methods of monitoring levels of contaminants and their effects on biota. Biomonitoring is a system of regular observations that provides information on the state of the environment by using living organisms as indicators for pollution status. The information is useful in predicting future changes and consequently managing the problems of environmental control. This theoretical basis (Murín and Mičieta 1993 + 1998, Murín 1999, Mičieta and Murín 2002c, Mišík 2002, Mičieta et al. 2003a, Mišík and Solenská 2004) impelled us to prove that this method could be successfully employed in practical cases. Subsequent studies demonstrated the sensitivity of a selected wild and cultivated species tested by the abovementioned method.

The method was proved effective for industrial complexes in Slovakia generally (Mičieta 1993, Mičieta and Murín 1997, Kunová-Staples and Mičieta 2000, Mičieta and Murín 1993a, b, Solenská et al. 2003, Mičieta et al. 2004a, Mičieta et al. 2005, Solenská et al. 2005a, b, Mičieta 2009, Brutovská et al. 2011, Mičieta et al. 2011a) as well as in specific cases: a nickel factory (Uhríková and Mičieta 1995, Mičieta and Kunová 2000a), a Smelting foundry (Mičieta and Murín 1996a), an aluminium factory (Mičieta and Murín 1996b, c), the Ore mine area (Kunová-Staples and Mičieta 2002), a paper-mill factory (Mičieta 1992a, Mičieta et al. 2002a, d, Mičieta et al. 2003a) and a chemical factory (Mičieta et al. 2013).

The genotoxicity of PAH evaluated by native flora came under our analysis too (Mičieta et al. 1998).

Detection of the ecogenotoxicity impact of agro-technologies on native vegetation was reported over a long period of time in several of our studies (Mičieta 1987, Mičieta 1988, Mičieta 1989, Mičieta et al. 2002c, Dienes et al. 2009, Mičieta et al. 2009a, Dienes et al. 2011) with the aim of adding some cultivated plants into our list of suitable bio-indicators (Mičieta and Murín 1996a).

Water is a natural medium for transporting potentially dangerous chemicals and for this reason it is an obvious object of our studies. Since 1980, we have been primarily focused on the Váh and Danube rivers (Murín et al. 1980, Murín 1984 + 1993, Murín et al. 2004a, Svetková et al. 2005, Solenská et al. 2007).

City environment is generally presumed to be unhealthy due to higher pollution, which was also reported in our studies (Mišík et al. 2003, Mišík and Mičieta 2006, Mišík et al. 2006, Solenská et al. 2006, Chrenová and Mičieta 2007, Mišík et al. 2007, Mičieta et al. 2007, Chrenová et al. 2010, Dušička et al. 2013).

Special attention was aimed at the environmental genotoxicity in the vicinity of a waste incinerator within Bratislava, the capital city of Slovakia (Mičieta et al. 1999a, b), and its industrial dumps (Mičieta and Murín 1998a, Mičieta et al. 1999, Murín et al. 1999, Solenská et al. 2006, Mičieta et al. 2010a, b, Sámelová et al. 2011). We have shown also the examples of how to re-cultivate them (Mičieta et al. 2002c, Dienes et al. 2009).

Demonstration of the general usefulness of this method has been shown by a collaborative study with Kuwait University, focused on the crude oil pollution caused purposely by the Iraq Army during the war in Kuwait (Murín and Malallah 1994, Malallah et al. 1997). Later, we took an interest in the environmental effect of normal production of an oil refinery (Mišík et al. 2007, Mičieta et al. 2008a).

The study of radioactive spots in the vicinity of the nuclear power station Jaslovské Bohunice was completed in cooperation with the University of Vienna, Austria (Mičieta and Murín 1994a, Murín and Mičieta 1996, Murín et al. 1996, Mičieta et al. 2006a, b, Mičieta and Murín 2007, Murín et al. 2007, Murín and Mičieta 2009b). This field of our interest was later enlarged by a parallel test of phytotoxicity and mutagenicity of spots contaminated by low-radioactive bombs used by NATO in Serbia (Mičieta et al. 2010c) and examples of radio-resistance by experiments with *Vicia faba* L. (Murín and Mičieta 2017a).

Also promising are results from retrospective biomonitoring of polluted environment by means of herbal samples as a unique method conceived by our laboratory (Mičieta 1998, Mičieta and Murín 1994b + 1996d + 1999).

The addition of tree and shrub species has expanded our list of suitable plant bioindicators (Mičieta and Murín 1995 + 1996a, b, c + 1997b, c, d + 1998b, c, d, Mičieta and Dušička 2000, Mičieta and Murín 2000, Murín and Mičieta 2000, Mičieta et al. 2000, Mičieta et al. 2006c, Pisarčíková et al. 2006, Mičieta et al. 2011c).

The effect of heavy traffic on the environment has been our focus for several years (Solenská et al. 2005c, Šípková and Mičieta 2005, Mišík and Mičieta 2006, Mičieta et al. 2008b, Mičieta et al. 2009b, Mičieta et al. 2011a, b, Mičieta and Gregušková 2013).

With all this knowledge, we are able to compare the sensitivity of invasive and indicative plant species as well as determine their other specific characterization for an eco-genotoxic deterioration (Mičieta et al. 2000, Dienes et al. 2010).

For future studies, we would like to use plant-specific locus systems in tests of genotoxicity (Brutovská et al. 2013).

All these findings allow us to offer practical outcomes in the strategy of phytoremediation (Mičieta et al. 2002b, Mičieta and Murín 2004, Dienes et al. 2009) and draw general conclusions about the unexpectedly high adaptive ability of living organisms to environmental stress (Mičieta1992a, Mičieta et al. 2002b, Murín 2002, Murín and Mičieta 2008a, b, c + 2009a + 2010, Mičieta and Murín 2014, Mičieta et al. 2017).

We do appreciate that this method was also adopted by other colleagues from the very beginning (Regis 1999, Regis and Lagunzad 2002, Regis et al. 2006, Solenská 2005, Samecka-Cymerman et al. 2012, Chropeňováet al. 2016b, Kandziora-Ciupa et al. 2016) and continues to be studied until the present day (Słomka et al. 2012 + 2014).

*Chapter 2*

# MATERIALS AND METHODS

## 2.1. ADVANCES OF "*IN SITU*" ENVIRONMENTAL STUDIES USING HIGHER PLANTS

1. Plants naturally cover all fields of our interests during the whole vegetation period.
2. There is no need to transport them to the place of an environmental accident with an obvious hazard of their stress and lack of adaptability that will affect the obtained results.
3. In the case of long-time growing plants, we may monitor them through their whole history and thus refer to their normal conditions in comparison with acute environmental pollution.
4. In case of a chronic environmental condition we may investigate their ability for an adaptation to the particular type of pollution.
5. In case of acute environmental pollution they cannot escape from the place as animals can.
6. Evaluation of their condition is easy and does not require an expensive investment.

7. The most effective markers are pollen grains that are easy to collect and their condition can be easily detected. They should also persist in good condition for evaluation for a very long time.

Operativeness and the outlooks of bioindication of mutagenic effects of the polluted environment by vascular plants lies in the possibility of long-term observations and their suitability for acute cases of pollution when it is impossible to prepare experimental conditions beforehand. The quoted methodological procedures of bioindication by the species of the local flora take into consideration all real environmental components, the nutrient flow, population dynamics, etc. Use of these plants facilitates detection of the effects and influences, which cannot be detected otherwise. The results can be extrapolated directly to humans assessing the risk for them. Procedures are operative, simple and financially accessible. They are suitable for long-term observations, as well as accidents of environmental pollution. These can be used for the assessment of the persisting old loads caused by acute exposition, and for the evaluation and comparison of the local and regional impacts.

## 2.2. Basic Criteria for a Selection of Plant Bioindicators

Apart from the pollen viability the impact of pollution is apparent as well on morphology of the pollen (Mičieta and Murín 1998a, b). Pollen morphology is one of the criteria for evaluation of pollen abortivity. The pollen abortivity test was used to evaluate the genotoxic impact of the environmental pollution on plants. This method is based on the work of Mulcahy (1981) and was used several times to evaluate the genotoxic potential of the environment in our reports mentioned above.

In the course of developing a method for "*in situ*" bioindication we observed localities with various types of predominate environmental risk factors which have a genotoxic effect, e.g., PCBs, VOCs, heavy metals, radionuclides, pesticides and heavy traffic air pollution by $SO_2$ and $NO_x$. These localities were in industrial complexes or heavy polluted urban areas, e.g., in the vicinity of the Bratislava Incinerator, the refinery in Bratislava, the nuclear power station in Jaslovské Bohunice, the paper mill in Ružomberok, the aluminum factory in Žiar nad Hronom, the nickel factory in Sereď and the ore mine areas in Banská Štiavnica and Pezinok, etc. (Uhríková and Mičieta 1995, Mičieta and Murín 1997a, b, Mičieta and Murín 1998a, b, Mičieta et al. 1998 + 1999, Kunová 1999, Mičieta et al. 2002a, d).

The basic criteria for the selection of plants as bioindicators using pollen grains as markers of polluted environment were described by Murín (1987) and later completed (Murín 1995). They are:

1. The selected species must be diploid (with pollen grains haploid), to avoid a spontaneous high pollen abortion which is common in polyploids.
2. The plants should produce well-developed and viable pollen grains under standard climatic conditions (spontaneous pollen abortion less than 5%).
3. The plants should have a common distribution in country-side and also grow in urban and industrial habitats.
4. Blooming seasons of various plant species should be in some sequence to cover all seasons of the year: spring, summer, autumn and winter, respectively.
5. The determination of type of selected plant species in question should be easy and indubitable for a common practice.
6. It is advisable to select both terrestrial and aquatic plant species.
7. It is an advantage to also choose (if possible) plant species with persistent pollen tetrads to distinguish the physiological and genetic damages of the pollen grains more easily.

Basic list of first suitable plant bioindicators for Slovakia is shown in Table 1. Others will be presented in following chapters.

**Table 1. List of species of Slovak native flora selected as suitable bioindicators in order of their blooming periods**

| Species | S | H | B |
|---|---|---|---|
| *Corylus avellana* L. | 1.4 ± 0.3 | 7.1 ± 0.8 R | III-IV |
| *Taxus baccata* L. | 0.2 ± 0.2 | 3.3 ± 0.5 R | III-IV |
| *Cornus mas* L. | 2.7 ± 0.3 | 10.7 ± 1.5 R | III-IV |
| *Salix caprea* L. | 3.9 ± 1.1 | 20.4 ± 4.4 R | III-IV |
| *Populus alba* L. | 3.1 ± 1.2 | 7.7 ± 1.5 R | III-IV |
| *Bellis perennis* L. | 2.0 ± 0.4 | - | III-X |
| *Lamium purpureum* L. | 3.0 ± 1.0 | 3.4 ± 0.3 A | III-X |
| *Anthriscus silvestris* (L.) Hoffm. | 3.3 ± 0.9 | - | IV-VI |
| *Pinus sylvestris* L. | 3.1 ± 0.3 | 5.9 ± 0.1 S | V-VI |
| *Robinia pseudoacacia* L. | 4.0 ± 0.3 | 9.0 ± 1.2 A | V-VI |
| *Iris pseudoacorus* L. | 1.0 ± 0.3 | - | V-VII |
| *Cerinthe minor* L. | 0.3 ± 0.3 | 5.9 ± 0.6 A | V-VII |
| *Chelidonium majus* L. | 2.0 ± 0.3 | 5.0 ± 0.4 A | V-VIII |
| *Chamomilla recutita* (L.) Rauchert | 1.8 ± 0.3 | 9.9 ± 0.8 A | V-VIII |
| *Nuphar lutea* (L.) Sm. | 1.2 ± 0.4 | - | V-VIII |
| *Ballota nigra* L. | 2.0 ± 0.6 | 5.7 ± 1.6 N | VI-VII |
| *Pinus nigra* L. | 3.0 ± 0.3 | 7.6 ± 0.3 S | VI-VII |
| *Ligustrum vulgare* L. | 0.5 ± 0.2 | 0.9 ± 0.2 A | VI-VII |
| *Consolida regalis* S. F. Gray | 1.8 ± 0.4 | 2.1 ± 0.6 N | VI-VIII |
| *Chaerophyllum bulbosum* L. | 2.2 ± 0.3 | - | VI-VIII |
| *Typha angustifolia* L. | 5.0 ± 0.8 | - | VI-VIII |
| *Alisma plantago-aquatica* L. | 1.6 ± 0.4 | - | VI-IX |
| *Calystegia sepium* R.Br. | 1.6 ± 0.2 | 4.4 ± 0.6 A | VI-IX |
| *Carduus acanthoides* L. | 1.2 ± 0.2 | 3.9 ± 0.8 N | VI-IX |
| *Conium maculatum* L. | 1.0 ± 0.4 | - | VI-IX |
| *Chenopodium hybridum* L. | 0.2 ± 0.1 | 0.4 ± 0.1 N | VI-IX |
| *Melandrium pratense*(Rafn) Roehling | 2.0 ± 0.4 | - | VI-IX |
| *Melilotus alba* Medic. | 1.0 ± 0.3 | 7.1 ± 0.7 A | VI-IX |
| *Melilotus officinalis* (L.) Pallas | 1.4 ± 0.4 | 4.5 ± 0.4 A | VI-IX |
| *Raphanus raphanistrum* L. | 5.0 ± 0.8 | 16.5 ± 1.5 A | VI-IX |
| *Sagittaria sagittifolia* L. | 2.0 ± 0.7 | - | VI-IX |
| *Sinapis arvensis* L. | 4.4 ± 0.8 | 7.9 ± 0.7 A | VI-IX |
| *Stachys annua* (L.) L. | 3.9 ± 0.4 | 4.4 ± 0.5 N | VI-IX |
| *Calluna vulgaris* (L.) Hull | 4.9 ± 0.2 | 27.4 ± 6.1 A | VI-X |
| *Conyza canadensis* (L.) Cronq. | 0.1 ± 0.1 | 0.2 ± 0.1 N | VI-X |
| *Mercurialis annua* L. | 1.6 ± 0.4 | 3.9 ± 0.9 N | VI-X |

| Species | S | H | B |
|---|---|---|---|
| *Berteroa incana* (L.) DC. | 1.2 ± 0.3 | 2.5 ± 0.4 N | VI-XI |
| *Daucus carota* L. | 4.4 ± 1.2 | 4.9 ± 0.3 A | VI-XI |
| *Linaria vulgaris* Mill. | 4.6 ± 1.1 | 7.9 ± 0.4 N | VI-XI |
| *Trifolium pratense* L. | 3.8 ± 0.6 | 9.9 ± 0.9 A | VI-XI |
| Species | S | H | B |
| *Artemisia vulgaris* L. | 2.4 ± 0.7 | 2.6 ± 0.5 A | VII-VIII |
| *Pastinaca sativa* L. | 3.4 ± 0.4 | 3.5 ± 0.3 A | VII-VIII |
| *Kochia scoparia* (L.) Schrad. | 0.2 ± 0.2 | 0.4 ± 0.1 N | VII-IX |
| *Cichorium intibus* L. | 1.0 ± 0.2 | 1.2 ± 0.2 N | VII-X |
| *Picris hieracioides* L. | 0.4 ± 0.2 | 1.6 ± 0.4 N | VII-X |

Legend: S – spontaneous frequency [%] of abortiveness of pollen grains, H – highest observed frequency [%] of abortiveness of pollen grains in polluted environment from emissions of heavy industry, B = blooming period [months].

A = Aluminium factory, N = Nickel factory, Np = nuclear power station, R = oil refinery, S = Smelting foundry.

* Raphanus raphanistrum L., *Linaria vulgaris* Mill., *Trifolium pratense* L., *Cerinthe minor* L., *Lamium purpureum* L., *Chelidonium majus* L., *Melilotus alba* Medic., *Pastinaca sativa* L., *Daucus carota* L. (these plant species were tested near the nuclear power station Jaslovské Bohunice on sites with increased radiation up to 322 kBq/kg of $^{137}$Cs, but without significant impact on abortiveness of pollen grains).

## 2.3. Establishing and Improving a List of Suitable Bioindicators

Finally 44 species were found as suitable for further studies. In comparison to the first report (Murín 1987), which gave a basic and essential impetus for our subsequent practical studies, some data were revised (e.g., *Calluna vulgaris*) and a new species were added (e.g., *Carduus acanthoides*) or excluded (e.g., *Butomus umbellatus* L., because it had been found to be triploid in some populations of the Slovak flora, c.f. Krahulcová and Jarolímová 1993). A compendium of these selected species of Slovak flora and their blooming periods is shown in Table 1. From this list *Artemisia vulgaris, Calystegia sepium, Chelidonium majus, Daucus carota, Lamium purpureum, Matricaria chamomilla, Melilotus albus, Melilotus officinalis, Pastinaca sativa, Raphanus raphanistrum* and *Trifolium pratense*, grown in the vicinity of the aluminium factory in Žiar nad Hronom, were useful for assessment. Among the 11 analyzed species, no significant differences between the samples from polluted and control areas were observed with *Pastinaca* sativa, *Artemisia vulgaris, Daucus carota* and *Lamium purpureum*. This verifies necessity that if we would, for different reasons, choose only these indicator species, we could, by not adhering to the methodical principle, deduce distorted and mistaken conclusions. It does not mean that these species are not be sensitive to other types of pollution. For example in the same area of Žiar nad Hronom area we found evidence of an increase in pollen abortion by other indicators from our list, especially by *Calystegia sepium, Chelidonium majus, Matricaria chamomilla, Melilotus albus, Melilotus officinalis, Raphanus raphanistrum* and *Trifolium pratense*.

For the crop plant species particularly, all information about the selected cultivar, its agrochemical treatment during the growth period and the weather conditions during the whole vegetation period has to be considered for a possible influence on the final results (Sheoran and Saini 1996).

The broad spectra of plant species allow us to utilize all parts of plants suitable for the detection of environmental pollution. They are pollen grains/microspores, micronuclei in tetrads and chromosomal aberrations/ mitotic abnormalities in meioses and mitoses.

### 2.3.1. Pollen Grains/Microspores

Young and closed flowers and flower buds from a sufficient number of randomly selected individual plants (min. 10 individuals) are collected and fixed in a mixture of ethanol:acetic acid (3:1). After 24 hrs, the fixing solution is replaced by 75% ethanol and the samples are stored until they are further processed. The use of pollen as an indicator of mutagenicity and phytotoxicity must be precisely defined since the pollen has high sensitivity, which very readily responds to altered conditions – such as extreme temperature, aridity, deficiency in nutrition, etc. Therefore, in evaluating the induced toxicity within polluted areas it is necessary to recognize the natural abortivity of the pollen. Higher frequency of pollen abortivity is influenced by the concentrations of the pollutant emissions during the period of meiosis, which is different for each species. Therefore, each of the selected species has to be under screen for 3 seasons (shortly before its blooming period), to avoid possible influences of extreme weather conditions, such as drought or fluctuation of temperature.

### 2.3.2. Preparation of Slides for the Evaluation of Pollen Abortion

Young and closed flowers are removed from the fixing solution, washed and the anthers excised. The pollen grains are pushed out from each anther with a needle, using a slight pressure. The debris is removed from the slide. The pollen grains are stained with 0.5% aceto-carmine or 0.05% aniline blue in lactophenol (Darlington and La Cour 1976, Kearns and Inouye 1993). After penetrating the stain, grains were covered by cover glass and examined under a microscope (Murín 1987 and 1995, Mičieta and Murín 1996a).

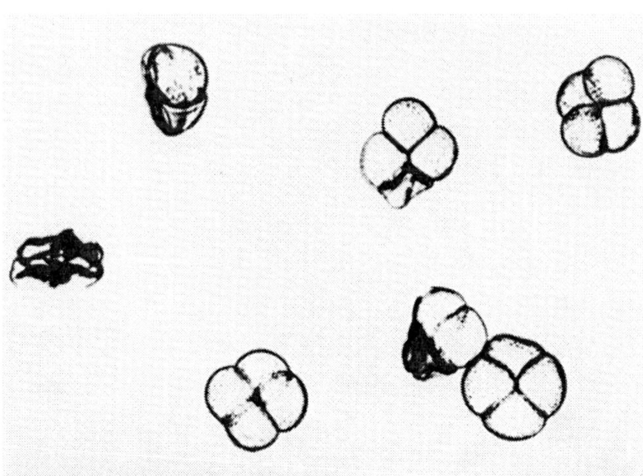

Figure 1 Pollen tetrads of *Typha latifolia* with 1, 2, 3 or all 4 abortive pollen grains.

In some plant species, the mature pollen grains occurred in tetrads and this characteristic offers a simple way of distinguishing between genetic and physiological toxicity of the pollutants in question. The abortion of one, two or three pollen grains in the pollen tetrad usually indicates genetic damage. Physiological damage usually affects whole flowers, anthers and all four pollen grains in tetrads. The genetic damage of all four pollen grains is so rare that, in the course of evaluation, including them among the physiological events will have no influence on the statistical significance (Mulcahy 1981) (Figure 1).

Pollen grains were evaluated for size, form and staining ability, with deviations considered as evidence for the lack of viability (abortion) or the possible occurrence of chromosomal aberrations (anomalies) (Figure 2).

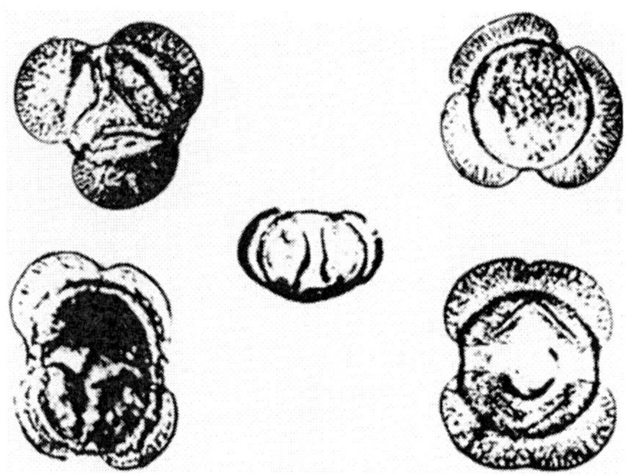

Figure 2. Sample of abnormal pollen grains of *Pinus nigra* L. from locality Lintich. Normal pollen grain in the centre.

### 2.3.3. Microspore Abortion in Tetrads

Flower buds containing early-stage tetrads of meiotic microspore mother cells are used for the evaluation of micronuclei. The four encased cells of the early tetrad usually remain together during slide preparation, while the later tetrad stage is inadequate for micronucleus scoring.

The young flower buds are macerated for 5-10 min in a mixture of ethanol:hydrochloric acid (1:1). The required materials have to be collected soon after finishing the tetragenese, when an exina of tetrads is not yet developed. Only under this condition is sufficient microscope evaluation available. After maceration, the flower buds are washed in distillated water. On the slide, the pollen tetrads are dissected from the flowers buds and anthers. All the debris should be removed from the slide. The tetrads are stained with acetic-orcein (Darlington and La Cour 1976). Staining could be supported by repeated heating over the flame or on a hot plate, and then by a gentle pressing under several layers of filter paper. About 3000 tetrads have to be evaluated from 10-20 flowers of different plants, e.g., 150-300 tetrads per slide.

### 2.3.4. Micronucleus Test

Young flower buds, containing early-stage tetrads of meiotic microspore mother cells are used for the evaluation of micronuclei. The required material must be collected soon after completing tetradogenese, when an exina is not yet developed. Sampled material should be fixed in an ethanol and acetic acid (3:1) mixture and after 24 hours replaced by 75% ethanol. The young flower buds are macerated for 5 – 10 minutes in the ethanol and hydrochloric acid (1:1) mixture and then carefully washed in distillated water. Then we have to dissect the bud, prepare anthers and release pollen mother cells from the anthers. The debris is removed from the slide and tetrads are stained with 1.0% acetic-orcein or acetic-carmin. The cover glass slide is applied and repeatedly heating over a flame of up to 60°C helps to facilitate staining. We then press gently on the cover slide with the hand. Several layers of filter paper are used as a cushion between the cover slide and hand to avoid breaking of the cover slide. Then we count micronucleus frequency under a microscope. At least 3000 tetrads from 10-20 flowers of different plants should be evaluated.

### 2.3.5. Mitoses

For the evaluation of chromosomal aberrations and mitotic abnormalities we have to sample the root tips cells or branches with young needles (in pines). Samples from five individuals per species are fixed in an ethanol:acetic acid (3:1) mixture. The fixative solution is replaced after 24 hrs by 75% ethanol. Samples are hydrolyzed for 10 min in the concentrated hydrochloric acid:ethanol (1:1) mixture at room temperature.

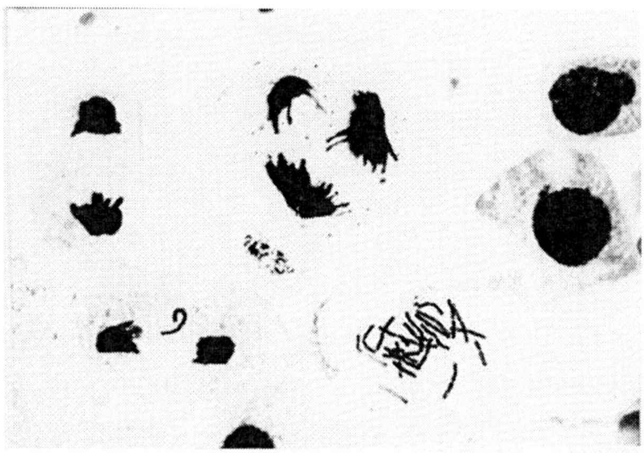

Figure 3. Samples of chromosomal aberrations and mitotic abnormalities in meristems of needles of *Pinus nigra* L. from locality Lintich hill (ana-telophase with fragment, c-mitosis, multipolar spindle).

Squash preparations are made from meristematic tissues. The chromosomes are stained in 1% of aceto-orcein. Chromosomal aberrations (C. A.) in ana-telophases (fragments, bridges, fragments and bridges) are distinguished from mitotic abnormalities (M. A. = lagging chromosomes, multipolar spindles and c-mitosis), because their different origins (sources of C. A. are affected chromosomes while M. A. are due to an injured mitotic apparatus, particularly mitotic spindle) (Figure 3). For each tested area 300 ana-telophases per species have to be scored.

### 2.3.6. Meioses

Preparation of slides for evaluation: For the analysis of meiosis, immature anthers were used. Slide preparation and evaluation are the same as above, a propion-orcein stain is used. We should identify and count the frequency of bridges, fragments and mitoclastic events (like multipolar spindles, c-mitosis and lagging chromosomes) in the anaphase of the 1st or 2nd meiotic fission in 300-500 anaphases cells from a minimum of 10 flowers of various individuals.

## 2.4. Tree and Shrub Species: Selection Criteria and Method of Sampling and Evaluation

A special attention in our studies was aimed to the tree and shrub species. The impair of the health status of forests is still a frequently discussed topic. This phenomenon is assumed to be a result of synergic effect of more stress factors. The pollutants act at various levels of living organisms and so the damage should be assessed at corresponding levels. Besides the physiological damage of the woody species, the damage of the genetic apparatus of the cells should be taken into consideration too.

Tree and shrub species are especially recommended for "*in situ*" indication of genotoxicity because of their long life span. Trees have been used as bioindicators of phytotoxicity, genotoxicity and mutagenity in the environment by several authors (Hool and Cairns 1995, Houghton 1995, Sparling 1995, Müller et al. 1997, Mičieta and Murín 1996 + 1997a, b + 1998a, b, Butorina et al. 2000a, b, Fleischer et al. 2000, Müller et al. 2000, Schlarbaum 2000, Butorina et al. 2002a, b, Kalaev et al. 2003, Aboal et al. 2004, Muratova and Sedel'nikova 2004, Gottardini et al. 2008, Carneiro et al. 2011). For example Bazhina et al. (2008) compared meiosis at microsporogenesis in *Abies sibirica* (Ledeb.) in natural populations and in an arboretum. Chropeňová et al. (2016a) tested impact of an artificial snowing along ski tracks to pine needles of nearby trees. Although usually used species are wooden or park trees, fruit trees may be used alternatively (Aroutiounian et al. 2004b).

Communities of the woody species cover almost 40% of the Slovak Republic landscape. Apart from the economic value, these communities present an important ecological element. Several woody species, utilized from the economical point of view, are exposed to additional stress factors like the chemical substances in the air with the genotoxic impact at the living organisms. Our studies deal with the impact of the various types of industrial pollution at the pollen abortivity of selected woody species. The pollen abortivity is regarded as a sensitive indicator of the genotoxic damage.

In the heavily polluted areas, the floristic spectra of arborescent species is usually drastically reduced and survival is primarily limited to introduced trees such as *P. sylvestris* and *P. nigra*. Their advantage as a bioindicators is also in the availability of their flowers. Usefulness of pines for ecotoxicological studies is well known for a long time (cf. Houston and Dochinger 1977).Male reproductive organs of plants are very sensitive to the environmental conditions. The impact of environmental stress factors on the process of microsporogenesis and pollen grain development can influence the biological properties of pollen. There are several reports that dealt with the changes of biological properties of pollen of trees in connection with the environmental pollution. Tretyakova and Noskova (2004) dealt with the impact of the environmental pollution on the pollen of *Pinus sylvestris* L. in urban agglomeration of city Krasnoyarsk. They evaluated morphological and biochemical characteristics of pollen. Apart from the dependence of the assessed parameters on the climatic factors they revealed the relationship between pollen properties and the level of the pollution. Oskolkov (1998) compared the pollen quality of *Pinus sylvestris* L. in dependence to the level of environmental pollution. Pukacki and Chałupka (2003) presented the decrease of viability of pollen and changes in the permeability of the cell membrane of *Pinus sylvestris* L. as the consequence of prolonging impact of atmospheric and soil pollution. Tretyakova and Bazhina (2000) have investigated the impact of the paper and cellulose industry on the vegetative and generative organs of *Abies sibirica* Ledeb. Apart from the others results they found out that the decrease of the fertility of the pollen is possibly connected with the decrease of diversity of this species in forest ecosystems.

Generative reproduction is the basis for the natural restoration and recovery of the gene pool of forest trees. It is important to detect genetic and physiological changes during the pollen ontogenesis and their influences at the pollen fertility, process of fertilization and seed quality and quantity. Many older papers (Houston and Dochinger 1977, Mulcahy 1981, Kappler and Kristen 1987, Wolters and Martens 1987, Ostrolucká et al. 1997, Ostrolucká and Fleischer 2003, Renzoni et al.1990, Renzoni and Viegi 1991, Seibert et al. 1992, Kristen and Kappler 1995, Mulcahy et al. 1996, Kristen 1997) referred to the differentiation of the generative organs of analyzed tree species. They described the early origins of defects at the pollen development level – anomalous pollen grains, reduction of pollen fertility degree, inhibition of pollen sac growth and seeds production.

The relationship between level of pollution and pollen viability can be connected with adaptation, which could occur after long-term exposition to the pollution. The problem of adaptation was reported by Holub and Ostrolucká (1993). These authors evaluated the adaptation of the pollen of *Pinus nigra* Arn. and *P. sylvestris* using the pollen germination at medium with increased concentrations of Pb. Ostrolucká et al. (1997) proved the adaptation of *P. nigra* to the atmospheric pollution of city agglomeration. Ostrolucká and Fleischer (2003) recorded the increase of the elongation rate of pollen tube from localities with increased concentrations of Pb in the soil. However, physiological damage of the individuals, from which the pollen was collected, may cause their death within few subsequent years. The adaptation to the chronic levels of pollution was mentioned in several our reports too (Mičieta and Murín 1999, Murín and Mičieta 2000a, b).

The acute damage of trees after short-term exposition of high level of pollutants causes leaf necrosis, tissue collapse and withering of whole trees. The chronic damage is a result of long-term exposure to low levels of pollutants. With organisms that have a long life span such as trees and shrubs, the detection of the chronic effects of pollutants can be observed. Some of these damages have no evident external signs. They are manifested as changes of physiological functions. Knowledge of these processes enables us to use trees and shrub species for long-term monitoring. Higher plants as comprehensive indicators in the environment generally comply with all the conditions and requests of the genotoxicity bioindication. For the ecological risk evaluation it is important to evaluate biota in the complete range of components in the environment in the condition "*in situ.*" Tree and shrub species as organisms with a long life span are convenient bioindicators of environmental mutagenity as well. In terms of standardization of our method and with the aim to broaden the set of indicators of the Slovak flora, we paid attention to selected unpolluted and polluted areas in Slovakia with an aim to present utilizable indication methods and a list of convenient tree and shrub indication species for a practical use. Our own studies are dated since 1995 (Mičieta and Murín 1995 + 1996 a + 1997a, b, c, d + 1998a, b, c, d, Mičieta and Dušička 2000, Mičieta and Murín 2000, Murín and Mičieta 2000a, b, Mičieta et al. 2000 + 2006a, Pisarčíková et al. 2006, Mičieta et al. 2008a, b + 2011a, b, c). Their results point at the efficiency and adequacy of the use of tree and shrub species as bioindicators of genotoxicity in the polluted environment, especially in the vicinity of industrial centres.

### 2.4.1. Selection Criteria for Tree and Shrub Species for Bioindication

The selection of convenient tree and shrub bioindicators species is oriented to dominant forest trees or common park species (cultivars) and allochthonous introduced species. As a material we can use closed flowers buds and microstrobiles at the beginning of pollen differentiation. Closed flowers are used to avoid the possible mixing of pollen with the pollen from another species and thus possibly following a wrong evaluation of samples.

A standardized procedure for genotoxicity indication by the microsporogenesis process of the tree and shrub species had not been prepared till our studies started. Evaluation of the genotoxic effect of pollutants is possible if using a method for the evaluation of chromosome and genome mutation in meristematic tissue. However, it is difficult to find a sufficient amount of material for cytogenetic analyses in the condition "*in situ*" (Müller et al. 1997). It is also possible to use analyses of seed germination from the various polluted areas (Bavcon et al. 1992).

There are many advantages for using tree and shrub species for the "*in situ*" genotoxicity evaluation (Mičieta and Murín 1996a):

- Possible indication of pollution or other risk factors in the complex setting of environmental components. It results from the integrative function of plants in the dynamical media.
- Tree and shrubs are organisms with a long life span and thereby accumulate all negative factors from the environment. Contrary to animals that naturally try to escape or avoid the contaminated environment, plants are constantly in the same location.
- Observation, indication and monitoring of genotoxic or mutagenic effect of risk factor or factors, residuums or metabolites are possible.
- High sensitivity of plant tests, which are given by the haploid state. All lethal mutations are manifested. We can calculate with mutability of 1,000 locuses.
- Abundance, which allows us to obtain an adequate sample set for competent quantification of the results in the real complex of all ecological factors and types of pollution.
- Indication of genotoxicity effects is available at all the structural and functional levels of genetic apparatus of cell.

For these reasons we regarded tree and shrub species as an important part of our list of suitable bioindicators (see Table 2).

#### *2.4.1.1. Criteria for the Selection of Shrubs and Tree Indicators*

The selection of shrubs and tree indicators has to fulfill specific criteria, which revert to works of Murín (1995), Mičieta and Murín (1996a):

- Use only diploid species constituting haploid pollen grains, the genes which are responsible for the normal development of pollen must be presented in one set of chromosomes, thus facilitating an immediate detection of their lesion and to avoid high pollen abortion common in polyploids.

- Selected trees and shrubs should produce well-developed and viable pollen grains under the normal climatic conditions and their pollen abortion cannot surpass 5% in the control locations.
- Use species currently occurring in the study area, which have a common distribution and which also grow at the secondary locations in wider environment (urban and industrial habitats).
- Choose various species, which have various blooming seasons. The blooming seasons of the various species should occur in sequence.
- The selected trees species should be easy to identify.
- When possible it is advantageous to choose plant species with pollen tetrads, which persist for a long-time. Tetrads are allowing us to easily distinguish the physiological and genetic damages of the pollen grains.

Although several reports deals with the changes of pollen viability in relationship to the industrial pollution, pollen responds sensitively as well to other environmental stress factors in response to the changes of the climatic conditions and they proved the relationship between pollen viability and changes of average temperatures in *P. sylvestris* (Parantainen and Pulkkinen 2002).For all species of genus *Pinus* the chromosome numbers are 2n = 24 and meiosis is generally regular (cf. Kumar et al. 1966). A detailed karyological study of *P. sylvestris* and *P. nigra* (Baranec 1979) showed that these species are fulfilling the criteria for bioindicators. Their spontaneous frequency of pollen grain abortion/anomalies from the control site did not exceed 5%.

#### *2.4.1.2. Prerequisites, Advantages and Limitations of the Method*

The prerequisites for a reliable evaluation of the impact of pollution using tree and shrub plant indicators are:

- Precise taxonomical identification of analyzed individuals.

- Picking of flower buds or microstrobules from at least 5-10 individuals, and from more branches of various age.
- Picking of control samples in the unpolluted locality at about the same time as the picking of the study samples.
- Fulfill all the requirements of GLP (good laboratory practice) for microscope analyses of samples (slide coding, etc.).
- Adequate knowledge of quality and quantity of risk factors.

We can state that this method is operative and its advantages are:

- Bioindication and evaluation of phytotoxic, genotoxic and mutagenic factors in the environment, differentiation of the local from regional impacts.
- Genotoxicity risk evaluation in a region and interpretation of its impact at the observed taxa and another living organisms including human.
- Indication of genotoxicity affects all structural and functional levels (gene, chromosome, genome) in the genetic apparatus of cells.
- Possibility to evaluate the individual adaptation response under chronic or repeated exposure.
- Monitoring of responses and the comparison of polluted and unpolluted localities.
- Simple economical procedure, standardized by the authors, does not need sterile condition, media for cultivation, preparation of experimental conditions, etc.

However, there are some limitations, which result from:

- Import of plants is masking the actual origin of planted individuals and their substratums. We recommend choosing native species to avoid introduced and invasive species.
- Short blooming periods of tree and shrub species.

- Tree and shrubs bloom only once per year and the amount of flowers may not be satisfactory.
- Sampling may be limited by the height of the individual tree.
- Sampling may be limited also by a limited number of individual target trees or shrubs in one location.

### 2.4.2. Sampling, Slide Preparation and Evaluation

For evaluation of genotoxicity we can use the following methods:

- Evaluation of the frequency of aberrant and anomalous microspores and tetrad analysis by convenient species for bioindication.
- Evaluation of clastogenic and mitoclastic effects in meiosis (micronucleus test).
- Evaluation of mitosis damage in the meristematic cells (root tips, leaf and flower stalks, needles of brachyblasts), indication of clastogenic and mitoclastic effects of environmental chemicals.

Pollen mother cells, microspore tetrads and microspores can be used to indicate genotoxic effects on the cell, organism, individual and population levels. It is possible to evaluate abortion of microspores and abortion of microspores in tetrad at the gene level (Mulcahy 1981), micronucleus's in microspores at the chromosome level (Mičieta and Murín 1997a), and frequency of anomalous microspores (unreduced 2n, 4n) at the genome level (Mičieta and Murín 1997b). Each of these levels has typical signs and needs a competent methodological procedure and a proper type of an organism.

There are five independent methods that can be used in combination for collecting more objective results.

### *2.4.2.1. Abortion of Microspores*

From chosen indicator species male generative organs, young and closed flowers or microstrobules at the beginning of their blooming period are collected. Multiple blooms from at least 5 individual trees or shrubs are collected from each location. The dispersion of individuals must be low (depending on the individual's size, abundance, dominance etc.). We recommend sampling locations along a line from the source of the pollution. This procedure requires knowledge about the source and type of pollution. The control samples from the unpolluted environments should be collected in the same time period using the same collecting method.

For example male strobili from *P. sylvestris* and from *R. pseudoacacia* were collected at the relevant sites. From each tree or shrub we examined at least 10 flowers. From these 10 flowers we must examine at least 3,000 microspores, usually 300 per slide. The basic criteria for evaluating microspores are their form (different shapes like normal developed microspores, non-developed, contracted microspores), their size (difference from the standard size specific for each species can be measured by micrometer) and their staining ability (reduced cell contents or un-stained). These deviations are considered as evidence for lack of viability. Microspores that are larger than their standard (e.g., non-reduced) are evaluated as anomalous.

Often we use the sampling line method when we evaluate pollen abortion of selected wild plant species at the localities along a line from the source of pollution at specified distances.

Using the obtained data we can calculate a reference index, which gives us information about the average microspores abortion of the samples from the polluted area higher in comparison to the control area. The reference index is important for the determination of spatial distribution of genotoxicity from the source of pollution and interpreting data in both time and space.

For details of preparation of slides see 2.2.2.

#### *2.4.2.2. Microspore Abortion in Tetrads*

In some plant species the mature microspores occurred in tetrads after the end of anatelophases. It gives us the ability to distinguish easily their genetic and physiological damage as described in a general chapter of Materials and methods (2.2.3.).

#### *2.4.2.3. Micronucleus Test*

This method is more or less informative because significant results can only be found in the heavy polluted areas. For details of preparation of slides see 2.2.4.

#### *2.4.2.4. Chromosomal Aberrations and Mitotic Abnormalities in Mitosis*

As a material for evaluation of chromosomal aberration and mitotic abnormalities branches with young needles, leaves or flower stems from at least 5 individuals per species are used.

It is difficult to find root tips, which are suitable for cytogenetic analyses in the condition "*in situ.*" Another possibility is using seeds from indicator species for seed germination.

For details of preparation of slides see 2.2.3.5.

#### *2.4.2.5. Chromosomal Aberrations and Mitotic Abnormalities in Meiosis*

The process for the preparation of material and slides for the evaluation of chromosomal aberrations in meiosis in microsporocytes (pollen mother cells) is the same as for the micronucleus test. For details of preparation of slides see 2.2.6.

### 2.4.3. A List of Bioindicators from Tree and Shrub Species of the Slovak Flora

After repeated analyses of the most of the diploid tree and shrub species of Slovak flora we were able to prepare a list of convenient indicators (Table 2).

**Table 2. Tree and shrub species as convenient bioindicators of the Slovak flora**

| **Species** | **Spontaneous abortivity** | **Max found abortivity** | **Chromosome number (x)** |
|---|---|---|---|
| *Abies alba* Mill. | 0.4 ± 0.2 | 22.5 ± 2.8* | 12 |
| *Acer platanoides* L. | 3.5 ± 0.5 | - | 13 |
| *Alnus glutinosa* (L.) Gaertn. | 1.8 ± 0.3 | 3.9 ± 0.6 | 14 |
| *Betula pendula* Roth. | 2.0 ± 0.5 | 7.1 ± 0.3 | 14 |
| *Celtis occidentalis* L. | 2.5 ± 0.3 | - | 10 |
| *Colutea arborescens* L. | 3.6 ± 0.4 | - | 8 |
| *Cornus mas* L. | 2.7 ± 0.3 | 10.7 ± 1.5* | 9 |
| *Corylus avellana* L. | 1.4 ± 0.3 | 9.0 ± 0.8* | 11 |
| *Fagus sylvatica* L. | 4.5 ± 0.5 | 9.4 ± 2.0 | 12 |
| *Juniperus communis* L. | 1.7 ± 0.3 | 4.3 ± 1.2 | 11 |
| *Juniperus virginiana* L. | 1.6 ± 0.2 | 5.6 ± 1.6 | 11 |
| *Larix decidua* Mill. | 2.3 ± 0.6 | 5.2 ± 0.5 | 12 |
| *Ligustrum vulgare* L. | 0.5 ± 0.2 | 0.9 ± 0.2 | 23 |
| *Negundo aceroides* Moench | 2.1 ± 0.3 | 8.2 ± 0.9* | 13 |
| *Picea abies* (L.) H. Karst. | 4.4 ± 1.2 | 18.6 ± 3.2 | 12 |
| *Pinus cembra* L. | 4.2 ± 0.4 | - | 12 |
| *Pinus nigra* Arn. | 3.2 ± 0.1 | 9.7 ± 0.6* | 12 |
| *Pinus mugo* Turra | 1.9 ± 0.3 | 5.4 ± 1.31<br>15.6 ± 2.22* | 12 |
| *Pinus sylvestris* L. | 3.6 ± 0.2 | 11.2 ± 1.5* | 12 |
| *Platycladus orientalis* (L.) Franco | 1.8 ± 0.3 | 6.9 ± 1.4* | 11 |
| *Populus alba* L. | 3.1 ± 1.2 | 7.7 ± 0.8 | 19 |
| *Populus nigra* L. | 4.7 ± 0.5 | 11.5 ± 1.6* | 19 |
| *Pseudotsuga menziesii* (Mirb.) Franco | 2.6 ± 0.3 | 3.7 ± 0.7 | 12 |
| *Robinia pseudoacacia* L. | 4.0 ± 0.3 | 34.2 ± 2.4* | 11 |
| *Salix caprea* L. | 3.9 ± 1.1 | 20.4 ± 4.4* | 19 |
| *Taxus baccata* L. | 0.2 ± 0.2 | 3.3 ± 0.5* | 12 |
| *Thuja occidentalis* L. | 1.4 ± 0.2 | 8.1 ± 0.8 | 11 |
| *Viburnum opulus* L. | 3.2 ± 0.3 | - | 9 |
| *Viburnum rhytidophyllum* Hemsl. | 3.0 ± 0.5 | 4.3 ± 1.3 | 9 |

Abortivity in %, 10 000 microspores per species evaluated) *$P \leq 0.05$.

From this list were excluded species which form hybridogenous complexes and are usually difficult to determine especially during the blooming period (like *Salix, Populus, Quercus, Fruxinus* sp.) and which show a high level of non-induced frequency of microspores abortivity. When we cannot find sufficient amounts of non-problem taxa species at the stand, we can use other taxa with pollen abortion at the control stand not higher than 5% after their precisely taxonomical determination. The special problem of using introduced species is the difficulty to precisely determine them taxonomically. Also, it is difficult to find sufficient amounts of individuals at the locality, and the control samples of the same cultivar etc. To comply with our method we can also successfully use park trees and shrubs for the genotoxicity indication in the urban territory.

The obtained data can provide valuable information about the impact of pollutants at the local and regional flora, especially with the results of the evaluation of wild plant species (Mičieta and Murín 1996a + 1997a).

The frequency of microspore abortion is influenced by the concentration of the pollutant emissions during the period of meiosis, which is different in each species. In terms of standardization of the method we observed seasonal variability (in course of 1990-2000), individual variability and variability between various flowers of one individual at the selected stands (Mičieta et al. 1999b).

In various populations from extremely polluted conditions were found different genetic structures and variability's. This refers to the adaptation and selection of genotypes, which benefited in tolerant plant species. A higher frequency of homozygotes was found in the group of sensitive trees and heterozygotes were found in the group of tolerant trees. A greater number of heterozygotes and genetic polymorphism results in a greater ability of individuals and populations to adapt and survive in polluted conditions (Prus-Glowacki and Novak-Browy 1992, Prus-Glowacki and Godzik 1993). If we know which are the adapted individuals and tolerant species, we can recommend species for its planting in urban areas and for reforestation of polluted areas.

## 2.5. Statistical Analysis

Statistical analysis between the samples from a control area and polluted areas is recommended to be conducted following the procedure of Amphlett and Delow (1984), which is based on the Poisson distribution.

This guarantees that the confidence level is at least $\alpha=0.05$ also for probabilities close to zero. For the data, an analysis of variance (ANOVA) and Wilcoxin sign-rank (one sided) test are also useful. All evaluations have to be done under blind conditions.

In several cases Kruskal-Wallis test was also on help (Theodorsson-Norheim 1986). For a comparison of the sensitivity of the herbaceous and woody species reference indexes were calculated. Reference index is defined as the ratio of pollen abortivity in impacted site and control site. Furthermore the results of pollen abortivity in localities were compared to the control locality using t-test for independent samples.

## 2.6. Usefulness of a Presented Method

Our long-term aim was to establish the original method of biomonitoring of polluted environment with wild plant and crop plant species for "*in situ*" conditions as an operative, effective and orientative front line tool in ecological and ecotoxicological studies. We would like to present this method as a test that could be used as a first step to recognize the current situation in an interested area/region. Afterwards, it allows us to decide if it is worth checking the genotoxic impact by another standard tests when the situation (described by results from this our test) requires it. This is maybe one of the main values of the method, which could thus serve for the broad practical use as a preliminary test.

*Chapter 3*

# Adaptations of Higher Plants to the Environmental Stress Caused by Heavy Metal Pollution of Soil in Industrial Complexes

## 3.1. Introduction

Industrial complexes were very well-known for their heavy pollution in the past, and therefore the aim of our study was to find a possible genotoxic effect on the samples of plant species. The connection between these factors was reported and plants as models for tests of a heavy metals recommended by Fargašová (2012). Sharma and Raju (1983) were among first to use vegetation around Hindustan polymers factory for a cytogenetic monitoring. Svetková et al. (2005) reported the use of higher plants for genotoxicity assessment of waste water polluted by heavy metals. Udosen et al. (2006) used Manihot utilissima to test an extractable heavy metals in soil within a municipal dumpsite. Ebong et al. (2007, 2008) used Carica papaya and Talinum triangulare for testing a heavy metal contents of municipal and rural dumpsite soils. Słomka et al. (2012) showed at *Viola tricolor* that heavy metals in soil affect reproductive processes more than morphological characters. This was shown in practice by Geras'kin et al. (2011) by means of geno-toxicity assay of sediment and water samples from the Upper Silesia post-mining areas, Poland, with an *Allium*-test. Feketeová et al. (2016) reported biological properties of extremely acidic cyanide-laced mining waste. Above all, for this chapter is important to note that Regis alone (1999), with Lagunzad (2002) and with other collaborators (2006) successfully used our method of testing of pollen grain's abortion as indicator of mercury pollution near gold mining sites in Philippines. Our laboratory was focused to industrial areas in Slovakia in several studies (Solenská et al. 2003, 2005 + 2005b).

Very interesting is fact that exist plant species that are doing well exactly in areas polluted by heavy metals as an obligate metallophytes (Bothe and Słomka 2017).

According to our method the pollen abortiveness of plants collected in such environment and finally identified as caused by mutagens and/or carcinogens will support the idea of pollen analyses as a suitable tool for a biomonitoring of polluted environment by means of wild plant species collected directly in the area of our interest. Special selected aspects for evaluation of phytotoxicity and genotoxicity of heavy metals within this method was established too (Kunová-Staples and Mičieta 2000, Mičieta et al. 2004b). Our method was tested on four examples of different forms of heavy metal pollution in Slovakia – in the vicinities of the Nickel Factory, the Smelting Foundry, an Aluminium Factory and the Ore mine area.

## 3.2. Materials and Methods

### 3.2.1. Areas of Testing

#### *3.2.1.1. The Nickel Factory in Sereď*

Sereď City is located 62 km North-East of the capitol city of Slovakia, Bratislava. The studied area belongs to the Podunajská nížina Lowland (126 m a.s.l), and it is close by the river Váh. The production of nickel was based on an ore imported from Albania. Its processing brought great amounts of smelting waste and consequently bulky heaps. The production of nickel resulted in huge amounts of industrial waste material consequently interfering with the natural character of the landscape, causing a constant deterioration of the ground waters and air quality in the vicinity of Sereď City. These negative effects on vegetation were reported by Zeleňáková and Holub (1985). The soil around these mountain-like heaps was polluted by the solid flow of dusts of the smelting waste containing heavy metals like Ni, Cr and Co. The accessible content of these elements is decisive for the plants. The extractible parts of the smelting waste were: Ni = 631 $mg.kg^{-1}$, Cr = 128 $mg.kg^{-1}$, Co = 185 $mg.kg^{-1}$ (Zeleňáková and Holub 1985). Washed corn leaves collected in the vicinity contained about 10-fold and in the distance of 4 000 m from the dump double amounts of Ni, Cr, and Co, respectively, as compared to the control (Zeleňáková and Holub 1985). Nickel production in Sereď City ended in 1993 (Mičieta and Kunová 2000).

#### *3.2.1.2. The Smelting Foundry*

Banská Štiavnica City, surrounded by woody hills, is situated in central Slovakia and has a long history of mining. The testing area was at Lintich Hill, with a slope of 300 m that descendent to the Smelting foundry, which produced emissions with elevated levels of Pb and Cu from 1783 until 1969.

Content of heavy metals in $mg.kg^{-1}$ (Holub et al. 1987):

- in soil: Pb = 5 500, Cu = 770, Zn = 720, As = 130, Ag = 35.6
- in several-years old needles of *Pinus nigra*: Pb = 109, Cu = 4.0, Zn = 61, As = 5.1, Ag = 0.0
- in one-year old needles *Pinus nigra*: Pb = 480, Cu = 9.0, Zn = 195, As = 4.8, Ag = 0.0
- in several-years old needles of *P. sylvestris*: Pb = 1080, Cu = 5.4, Zn = 93, As = 645, Ag = 7.0
- in one-year old needles *of P. sylvestris*: Pb = 2260, Cu = 6.6, Zn = 237, As =1689, Ag = 4.1.

PB content in pollen grains of both species was 9.54 – 10 (Holub and Ostrolucká 1993).

### *3.2.1.3. The Aluminium Factory*

The factory is located in the Žiarska kotlina basin in central Slovakia (Figure 4). The technological cycle includes not only the production of aluminium, but also of $Al_2O_3$ and an electrode paste. Aluminium is obtained by electrolysis of $Al_2O_3$ with the use of Söderberg electrodes. Yearly, 600,000 t of bauxite and coke and 300,000 t of brown coal (0.009% As and 29.55% of ash) was processed. More than 15,000 t of emissions were produced a year, with 10,000 t being gaseous. Emissions of $F^-$ were 800-1000 t per year, next is 42.5 t of As, 800 t of tar and phenols, 0.86 t of Hg and others (Maňkovská et al. 1988).

Figure 4. Location of Žiar nad Hronom, the monitored areas and photograph of the Aluminum factory.

The study site was approximately 1000 m from an Aluminium factory, with emissions containing extremely high concentrations of fluorides, $SO_2$, As, PCB, PAH as: fenantren = 2 $mg.kg^{-1}$, fluoranten = 3.3 $mg.kg^{-1}$, pyren = 1.7 $mg.kg^{-1}$, chryzen = 1.4 $mg.kg^{-1}$, perylen = 3,1 $mg.kg^{-1}$ (Mičieta et al. 1998).

- Content of metals in soil (Holub in litteris) in $mg.kg^{-1}$:
  - Al = 127 – 156, As = 40 – 160, Cr = 60 – 160
- Content of metals in dry plants (Holub in litteris) in $mg.kg^{-1}$:
  - Al = 405 – 1120, As = not detected, Cr = not detected

The composition and amount of these emissions was so extensive that a presumption of not only toxic but also mutagenic effects on living organisms is reasonable. This includes polycyclic aromatic hydrocarbons (PAH) that evaporate from the electrode paste in Söderberg process, consisting of coal tar and pitch (Maňkovská et al. 1988). Benzo(a)pyrene, an important carcinogen is a typical component of PAH. The territory within the radius of 10-15 km from the factory was visibly threatened. Located in a natural valley, with a frequent occurrence of atmospheric inversions, increased the concentrations of emissions containing extremely high concentrations of fluorides, $SO_2$, As, PCB and PAH.

#### *3.2.1.4. The Ore Area*

From the phytogeographical point of view, the site is at South-East slopes of Malé Karpaty Mts. (240 – 265 m a.s.l.), about 4 km North-West from the centre of Pezinok city (latitude 48°17′, longitude 17°16′).

Sludge and soil around the sludge bed is polluted by the heavy metals like As, Sb, Cr, Ni, Zn, V, Mn. Analyses by Letko et al. (1992) ascertained in the sludge ($mg.kg^{-1}$): As>1,000, Sb >1000, Cr = 500, Ni = 250, Zn = 100,V = 500, Mn = 1,000, Cu = 50, Pb = 10 and in the soil around the sludge bed ($mg.kg^{-1}$) As >1,000, Sb >1,000, Cr = 500, Ni = 250, Zn = 200, V = 500, Mn = 1,000, Cu = 100, Pb = 100. Pedological analyses (Holub and Zeleňáková 1985) of fluvial soils of Pezinok brook basin showed contents (ppm) of As = 33.4 – 118.6, Sb = 15.5 – 111.8, Ni 44.0 – 87.0, Zn = 113.0 – 195.0, V = 118.0 – 162.0, Cu = 43.0 – 99.0, Pb = 34.0 – 75.0 in the A-horizon. Chemical analyses of sludge samples (Holub and Zeleňáková 1985) used for testing showed contents ($mg.kg^{-1}$) of As = 6.5, Ni = 62.5, Zn = 21.9, Mn = 410.2, Cu = 61.0 and Pb = 87.5.

### 3.2.2. Collection of Samples

*Pinus sylvestris* L. and *Pinus nigra* Arn. were chosen for evaluation as bionidicators of polluted environments. These species are commonly spread in Slovakia and are used for reforestation of devastated areas nearby sources of industrial pollution. Also *R. pseudoacacia* and 16 other arborescent species were tested. These are wild plant species, common at these localities and their blooming period has a long duration. Trees were used in our study for a first time. For all species of genus *Pinus* the chromosome number is 2n = 24 (cf. Kumar et al. 1966). A detailed karyological study of *P. sylvestris* and *P. nigra* (Baranec 1979) showed that the species fulfill the criteria for bioindicators (see 2.2.). Their spontaneous frequency of pollen grain abortion/anomalies from the control site did not exceed 5%.

Flowers were collected in 10 day intervals during the blooming period from the middle of July until the middle of September in 1993 and 1999. Collected samples were fixed in an ethanol and acetic acid (3:1) mixture. This mixture was replaced by 75% ethanol after 24 hours.

As a first step the material from all available diploid tree and shrub species from the unpolluted areas (Záhorská nížina, Lowland – villages Veľké Leváre, Moravský sv. Ján and the Botanical Garden in Bratislava) had to be sampled to obtain a data about the spontaneous frequency of microspore abortivity. The content of heavy metals in the soil and also in plant material from there was found to be below the background level. After this preliminary evaluation, we widened our focus to include polluted areas.

#### *3.2.2.1. The Nickel Factory*

Species of the local flora surrounding the dump were selected according to the already established criteria (Murín 1987, Mičieta and Murín 1996 a).

Pollen abortivity was evaluated at the proximity of the Nickel factory and on the transect line from the Nickel factory to the village Vlčkovce with a gradient of 500 m (Locality 1-6, Table 4).

Samples were collected in 1993 as one of our first studies (Uhríková and Mičieta 1995) and sampling was repeated then during practical exercises of our students in 2013. It gave us possibility to compare the effects of the long-term pollution by heavy metals in the same area within the range of twenty years.

#### *3.2.2.2. The Smelting Foundry*

*Pinus sylvestris* and *Pinus nigra* trees were artificially introduced to this area approximately 30 years before our testing. Male strobili of *P. sylvestris* and *P. nigra* from the three different areas were collected. In all areas ten tips of the young branches with microstrobules from five individuals from both species were fixed in an ethanol:acetic acid (3:1) mixture. The fixation solution was replaced by 75% ethanol after 24 hrs. The sampling was provided in year 1993 and again in 1996.

#### *3.2.2.3. The Aluminium Factory*

As in previous case the same tree species *P. sylvestris* and *P. nigra* were used as bioindicators.

The samples from the vicinity of an Aluminum factory and control samples of *Calluna vulgaris* from an unpolluted environment were collected at the same time from these localities:

- *Control #1*. Záhorská nížina Lowland is situated 194 km South-West of Žiar nad Hronom city. It is an area without any heavy industry.
- *Control #2*. This the area of Vysoké Tatry Mts. is situated 164 km North-East of Žiar nad Hronom. It is a tourist resort promoted as a climatic Spa with special aim to the healing power of mountain air.
- *Control #3*. Žiar nad Hronom city, the Aluminum factory.

#### *3.2.2.4. The Ore Mine Area*

Young flowers were collected in the surroundings of the sludge bed, from dumps and from a meadow approximately 100 meters opposite the sludge bed. Unpolluted localities in Záhorie were used as the control areas, as in a previous study. We also wanted to compare the pollen abortivity of native and implanted plant species, but the implantation of *Tradescantia paludosa* in the "*in situ*" conditions was unsuccessful (Kunová-Staples and Mičieta 2002).

In the conditions "*in vivo*" we tested solutions prepared from after-flotation sludge and distilled water in various ratios, water from Pezinok brook and mining water. Distilled water was used as the negative control and maleic hydrazide as the positive control.

We used flower buds that contain the early tetrad stage of meiotic pollen mother cells of *Tradescantia paludosa* according to Mišík et al. (2011). The tested solutions were applied directly through the flowers (evaluation after 24 hours) and indirectly through stems (evaluation after 72 hours).

### 3.2.3. Evaluation of Samples

Evaluation was performed according to the method described in Chapter 2. Details of the method were described by Murín (1987 + 1995) and optimized by Mičieta and Murín (1997b).

#### *3.2.3.1. The Nickel Factory*

We evaluated 100 pollen grains of every collected flower, 3,000 pollen grains or tetrads of each sample.

#### *3.2.3.2. The Smelting Foundry*

From each area, 3,000 pollen grains and 300 ana-telophases were evaluated per species.

In meiosis ana-telophases, the frequency of bridges, fragments and mitotic abnormalities (lagging chromosomes, multipolar spindles and c-mitosis) were determined from the meristems of needles. This technique is unique for these tests therefore examples of abnormalities found are showed at Figure 3 (see back in text).

The phenomenon of an increased frequency of unreduced pollen grains indicated a karyoclastic effect of emissions on these species when grown in proximity to heavily polluted areas (Figure 2, see back in text).

#### *3.2.3.3. The Aluminium Factory*

Evaluation of abortions of the pollen grains and chromosomal aberrations in meristems cells of pinuses was the same as above (3.3.2.).

For *Calluna vulgaris y*oung and closed flowers were removed from the fixing solution, washed and the anthers excised. The pollen grains were released out from each anther with a needle, using a slight pressure. The debris was removed from the slide. The pollen grains were stained with 0.5% aceto-carmine or 0.05% aniline blue in lactophenol (Kearns and Inouye 1993).

#### *3.2.3.4. The Ore Mine Area*

Abortivity was determined from 3,000 pollen grains for each sample.

We also scored the frequency of micronuclei of 1,500 tetrads from 10 flower buds for each sample. The same testing solution was used for the evaluation of abortivity of *Tradescantia* pollen grains.

For the Parallel test of phytotoxicity and mutagenicity (Murín 1984) chromosomal aberrations (fragments, bridges, fragments and bridges) were determined in 300 ana-telophases for each tested sample of mitosis of *Vicia sativa.*

### 3.2.4. The Selection of Suitable Species

Species like *Ballota nigra, Berteroa incana, Carduus acanthoides, Chenopodium hybridum, Cichorium intibus, Consolida regalis, Conyza canadensis, Daucus carota, Echium italicum, Kochia scoparia, Lamium purpureum, Linaria vulgaris, Mercurialis annua, Picris hieracioides, Stachys annua, Trifolium pratense* and *Typhalatifolia* has grown in tested areas and were useful for our analysis. For increased abortion of pollen grains, significant differences were observed in the following species:

*Ballota nigra, Echium italicum, Mercurialis annua, Picris hieracioides, Trifolium pratense* and *Typha latifolia*. In comparison to the first list of suitable species for "*in situ*" bioindication of genotoxicity (Murín 1987), trees were used at the first time as an initial assessment. The reason was simply the fact that around the Smelting foundry in Banská Štiavnica, only *P.sylvestris, P. nigra, Robinia pseudoacacia* and *Betula pendula* Roth. were able to survive. *Agrostis tenuis* Sibth. (fragmental monocenosis) and *Rumex acetosella* (sporadically) were also concurrently present in the area. We found *P. sylvestris* and *P. nigra* also in the vicinity of the Aluminium factory in Žiar nad Hronom.

## 3.3. Results and Discussion

### 3.3.1. The Nickel Factory

The results expressed in the frequency of occurrence of abortive pollen grains in collected samples are presented in Table 3. They univocally point at, regarding the significant assessment of the phytotoxicity and genotoxicity, the necessity of a selection of wider spectrum of indicators in the given locality. Out of 20 analyzed species only 9 belong to the recommended group of Murín (1987). The analysis of control samples of diploid species comprising *Kochia scoparia*, *Echium italicum*, *Consolida regalis*, *Carduus acanthoides*, *Picris hieracioides*, *Sonchus asper subsp. Asper*, *Stachys annua*, *Chenopodium hybridum*, *Conyza canadensis*, *Lactuca serriola*, *Atriplex tatarica*, shows that these species are also suitable bioindicators of the genotoxicity and may contribute to the existing group (Mičieta and Kunová 2000b). In 12 analyzed species, no significant difference in frequency of abortive pollen grains was found as compared to the control. Remaining species more or less imply the tendency of an increased abortion, while a clearly significant increase was observed only in the following species: *Ballota nigra*, *Carduus acanthoides*, *Echium italicum*, *Mercurialis annua* and *Trifolium pratense*. The pollen in *Typha latifolia* persists in tetrads, thus allowing to distinguish the genetically and physiologically conditioned abortion of pollen (Mulcahy 1981, Murín 1987).

We did not observed any tendency of increased abortion of pollen grains in tetrads in the sample collected near the heap. It is interesting that the pollen abortion is higher in the control of the species *Lactuca serriola* and *Atriplex tatarica* than in the samples from the exposed localities. At the assessment of the induced toxicity and eventually the genotoxicity of the polluted environment it is necessary to differentiate the natural abortion of pollen grains. A possible adaptation, eventually tolerance of the population, especially after a long lasting chronic or cyclic exposition cannot be neglected (Mičieta 1992 b, Chudzińska et al. 2015a).

**Table 3. Frequency of abortive pollen grains of native indication species of plants**

| Species | Control | Nickel factory | Control | Nickel factory |
|---|---|---|---|---|
| | 1993 | | 2013 | |
| *Atriplex tatarica* | 0.61 ± 0.2 | 0.20 ± 0.1 | 0.85 ± 0.2 | 1.24 ± 0.2 |
| *Ballota nigra* | 1.35 ± 0.3 | 5.71 ± 1.6 ++ | 1.31 ± 0.3 | 3.20± 0.4 ++ |
| *Berteroa incana* | 1.75 ± 0.3 | 2.45 ± 0.4 | 2.25 ± 0.2 | 2.01 ± 0.3 |
| *Carduus acanthoides* | 1.15 ± 0.2 | 3.95 ± 0.8 ++ | 1.00 ± 0.2 | 3.45 ± 0.5 ++ |
| *Cichorium intybus* | 0.30 ± 0.1 | 0.70 ± 0.1 | 1.15 ± 0.2 | 2.04 ± 0.2 |
| *Consolida regalis* | 1.75 ± 0.4 | 2.10 ± 0.6 | 1.70 ± 0.3 | 4.26 ± 0.8 |
| *Conyza canadensis* | 0.10 ± 0.1 | 0.13 ± 0.1 | 0.43 ± 0,2 | 0.80 ± 0.2 |
| *Daucus carota* | 2.90 ± 0.3 | 3.45 ± 0.6 | 4.40 ± 1.2 | 4.53 ± 0.5 |
| *Echium italicum* | 1.30 ± 0.3 | 5.05 ± 0.7 ++ | 1.60 ± 0.6 | 4.05 ± 0.7 ++ |
| *Chenopodium hybridum* | 0.23 ± 0.1 | 0.40 ± 0.1 | 0.46 ± 0.1 | 0.91 ± 0.4 |
| *Kochia scoparia* | 0.36 ± 0.1 | 0.40 ± 0.1 | 1.20 ± 0.2 | 1.70 ± 0.3 |
| *Lactuca serriola* | 0.86 ± 0.2 | 0.43 ± 0.1 | 1.51 ± 0.2 | 2.48 ± 0.3 |
| *Lamium purpureum* | 2.65 ± 0.3 | 3.90 ± 0.4 + | 2.82 ± 0.3 | 3.22 ± 0.2 |
| *Linaria vulgaris* | 1.25 ± 0.3 | 1.95 ± 0.3 | 4.50 ± 0.2 | 6.57 ± 2.2 |
| *Mercurialis annua* | 0.66 ± 0.2 | 3.83 ± 0.9 ++ | 1,03 ± 0.7 | 4.46 ± 1.2 ++ |
| *Picris hieracioides* | 0.43 ± 0.1 | 1.60 ± 0.4 ++ | 0.62 ± 0.1 | 1.44 ± 0.2 ++ |
| *Sonchus asper* subsp. Asper | 0.46 ± 0.2 | 1.20 ± 0.4 + | 0.84 ± 0.3 | 1.45 ± 0.3 |
| *Stachys annua* | 3.90 ± 0.4 | 4.33 ± 0.5 | 3.83 ± 0.6 | 3.95 ± 0.4 |
| *Trifolium pratense* | 1.85 ± 0.4 | 4.15 ± 0.6 ++ | 2.15 ± 0.3 | 3.33 ± 0.2 ++ |
| *Typha latifolia* | 0.10 ± 0.1 | 0.10 ± 0.1 | 1,01 ± 0.2 | 1,02 ± 0.1 |

++ $P \leq 0.01$.

The reliable interpretation of the results also requires an exact determination of the taxa (e.g., *Atriplex tatarica, Kochia scoparia*). In principle, a 20-year time-gap between sampling caused a slightly decreased effect in abortivity of pollen grains of most sensitive species used.

We were also interested in testing the transportation of pollution from its source by air (the Nickel factory) towards the nearby village Vlčkovce using another set of plant bioindicators (see Table 4). To our surprise, there was not a clear decrease of its impact by a distance (probably that distances were not long enough?) and for the sensitive species, as well as for lesser ones, the impact at the first locality remained almost at the same level till the last one. This is also confirmed by Induction index for species and Induction index for localities.

**Table 4. Pollen abortivity at the transect line from the Nickel factory to Vlčkovce. Gradient between localities is 500 m**

| Indication species | Average pollen abortivity in % | | | | | | Induction index for species |
|---|---|---|---|---|---|---|---|
| | Loc. 1 | Loc. 2 | Loc. 3 | Loc. 4 | Loc. 5 | Loc. 6 | |
| *Artemisia vulgaris* | 1.7 | 1.9 | 1.9 | 2.0 | 2.0 | 1.9 | 1.5 |
| *Carduus acanthoides* | 1.9 | 1.8 | 1.9 | 1.9 | 1.8 | 1.8 | 1.4 |
| *Daucus carota* | 4.7 | 4.8 | 4.9 | 4.5 | 4.9 | 5.4 | 1.1 |
| *Kochia scoparia* | 1.7 | 1.7 | 2.3 | 1.7 | 1.7 | 1.3 | 1.4 |
| *Linaria vulgaris* | 6.5 | 7.2 | 8.1 | 6.5 | 7.5 | 6.1 | 1.4 |
| *Chenopodium hybridum* | 0.9 | 0.8 | 1.2 | 0.9 | 0.8 | 0.8 | 1.9 |
| *Tanacetum vulgare* | 3.2 | 3.2 | 2.9 | 3.0 | 3.1 | 2.2 | 1.4 |
| *Datura stramonium* | 1.6 | 2.0 | 2.5 | 1.6 | 2.2 | 1.8 | 1.2 |
| Induction index for loc. | 2.8 | 2.9 | 3.2 | 2.8 | 3.0 | 2.7 | X |

### 3.3.2. The Smelting Foundry and the Aluminium Factory

Usefulness of pines for ecotoxicological studies is well known for a long time (cf. Houston and Dochinger 1977). In the heavily polluted areas, the floristic spectrum of arborescent species is usually drastically reduced and survival is primarily limited to introduced trees, such as *P. sylvestris* and *P. nigra*. Their advantage as bioindicators is also in the availability of their flowers. Usefulness of pollen grains from tree species as bioindicators of environmental pollution is twofold. Pollen grains are sensitive, e.g., exhibit morphological changes to emissions, and pollen grains provide large numbers of phenotypes that can be conveniently evaluated with a microscope (Mulcahy 1981). Additionally, the use of trees as bioindicators also provides continuous observations of the effects of emissions over a longer period of time. By using long-living bioindicator species, it is possible that adaptive responses to a chronic exposure can be observed. The low frequency of unreduced pollen grains of *P. nigra* in samples from the Smelting foundry, and a higher germination rate of pollen of both species over the control when cultured in media with high Pb concentration could be an evidence of the adaptive response (Holub and Ostrolucká 1993). Interesting results for adaptation mechanisms of *Pinus sylvestris* L. in industrial areas was reported by Chudzińska et al. (2015a, b). Usefulness of trees as bioindicators is summarized in monograph edited by Borzan and Schlarbaum (1997).

The comparison of samples from the control area with two polluted areas showed a significant effect of emissions on the frequency of abortive pollen grains (Figure 5). There was a differential sensitivity between *P. sylvestris* and *P. nigra* in the polluted areas (Table 5). Proximal to the Aluminium factory, the highest frequency of abortive pollen grains was observed in samples of *P. nigra*, in comparison with the control samples of *P. nigra* (2.43x higher) and 1.70x higher in comparison with the samples of *P. sylvestris*. In contrast, a higher observed frequency of abortive pollen grains was from *P. sylvestris* proximal to the Smelting foundry. These species had 1.93x and 1.17x more abortive pollen grains in comparison with the control samples of *P. sylvestris* and *P. nigra* from the same polluted area, respectively. The differences between the samples from the polluted areas and the control area were significant with a low level of standard deviation.

In addition to abortive pollen grains, pollen grains with diameters approximately 1.33x wider than control pollen grains were found in the polluted areas. For example in *Pinus sylvestris* the difference between normal and abnormal pollen grains per 1000 evaluated pollen grains was: d = 4.77 ± 0.8 μm vs. 5.91 ± 0.6 μm, r = 2.38 ± 0,4 μm vs. 2.95 ± 0.3 μm, which leads to evident difference in V = 567.98 $\mu m^3$ vs. 1080.29 $\mu m^3$. The larger pollen grains also had a higher number (3-4) of air bags (sacculus aereus) or a continual bag in the equatorial level (aequator) appeared. These pollen grains were regarded as unreduced, diploid pollen grains (Figure 5 in 3.3.2.).

The pollen grains of nearly the same shape were already observed in *Pinus pinea* L. collected from Pisa town centre affected mainly by pollution of Pb, S and $O_3$ (photochemical smog) (Cela Renzoni et al. 1990). The same studies were done in Slovakia (Chropeňová et al. 2016) and Poland (Kandziora-Ciupa et al. 2016) with alternative species *Pinus mugo* Turra and *Pinus sylvestris* L. Our results show that frequency of unreduced pollen grains of *P. nigra* and *P. sylvestris* was analogic with the species observed for abortive pollen grains in above mentioned studies. Around the Aluminium factory the frequency of unreduced pollen grains in samples of *P. nigra* was 4.86x higher than in the control samples and 1.32x higher than in *P. sylvestris*. Conversely, the highest frequency of unreduced pollen grains around the Smelting foundry was in the samples of *P. sylvestris* – in comparison with the control samples of *P. sylvestris* the frequency was 2.67x higher, and 1.85x higher when compared with the samples of *P. nigra* from same polluted area.

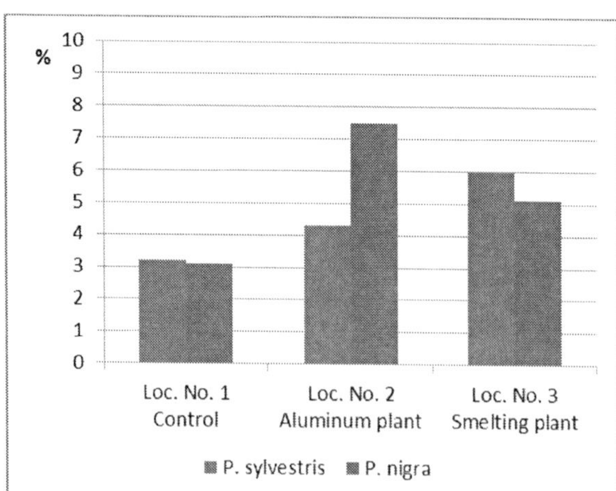


Figure 5. Frequency of abortive pollen grains in % in years 1993 and 1996. Control, the Aluminium factory and the Smelting Foundry.

Again, the differences between samples from polluted areas and the control area were significant with a low level of standard deviation, except in the samples of *P. nigra* collected in proximity to the Smelting foundry. Samples of chromosomal aberrations and mitotic abnormalities in meristems of needles of *Pinus nigra* L. as well as their frequency for both polluted areas are shown in Figure 6 and 7.

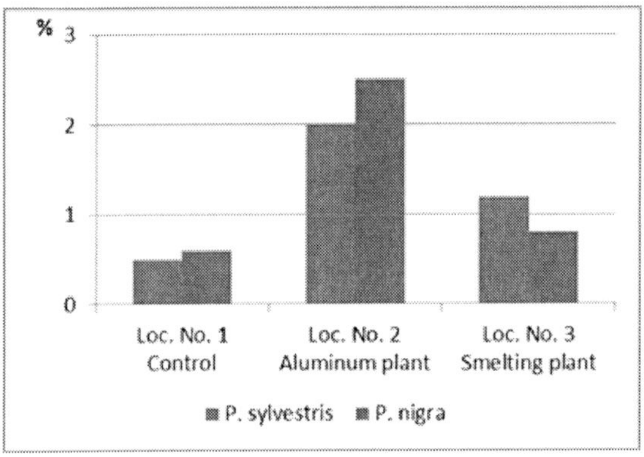


Figure 6. Frequency of abnormal pollen grains in % in year 1993 in control, the Aluminium factory and the Smelting Foundry.

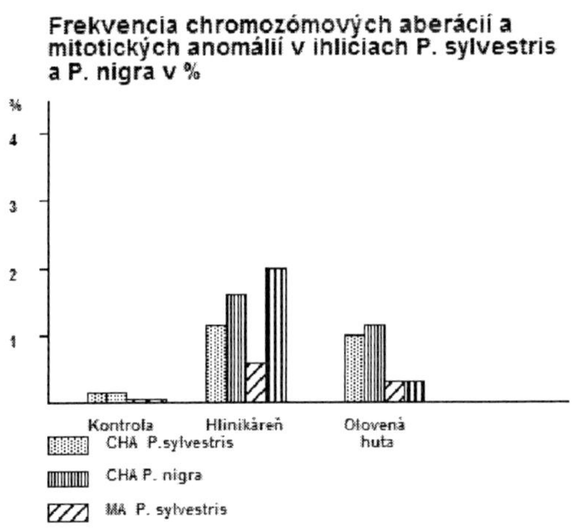


Figure 7. Frequency of chromosomal aberrations and mitotic abnormalities in meristems of needles of *Pinus sylvestris* L. and *Pinus nigra* L. in years 1993 and 1996. Control, the Aluminium factory (Hliníkáreň), the Smelting foundry (Olovená huta), CHA = chromosomal aberrations, MA = mitotic abnormalities.

**Table 5. Frequency of abnormal pollen grains of *Pinus sylvestris* L. and *Pinus nigra* L. at the Smelting foundry during years 2013 and 2014**

| | 2013 | 2014 |
|---|---|---|
| *Pinus sylvestris* L. | | |
| Control | 3.4% | 3.2% |
| Aberrations | 7.0% | 6.8% |
| Abnormal | 1.2% | 1.3% |
| *Pinus nigra* L. | | |
| Control | 2.8% | 2.7% |
| Aberrations | 6.1% | 5.7% |
| Abnormal | 0.5% | 0.6% |

To show the development in mutagenicity in the Smelting foundry till these days, here are the newest data obtained from our continual sampling and evaluation (Table 5).

### 3.3.3. The Ore Mine Area

Wild plant species enable us to evaluate the genotoxic effects of xenobiotics, which are present in the environment. Especially interesting are mining areas heavily affected by a human activities. However, we did not found significant differences in frequency and occurrence of abortiveness of pollen grains from the observed and control areas in 8 analyzed species (*Cichorium intybus* L., *Daucus carota* L., *Linaria vulgaris* Mill., *Melilotus albus* Medik., *Picris hieracioides* L., *Populus alba* L., *Salix caprea* L., *Trifolium pratense* L., Table 6). Adaptability, eventually tolerance of the population, especially after a long-term (chronic) exposition, is possible (Macnair 1987, Cibulka et al. 1991, Chudzińska et al. 2015).

*Tradescantia* micronuclei test is an efficient short-term "*in situ*"and "*in vivo*" bioassay that can provide the first alert for the presence of hazardous agents in air, soil or in drinking water (Ma 1982).

**Table 6. Frequency (%) of abortiveness of pollen grains of bioindicator species in the observed area**

| Species | Control | | Sludge bed | | Dumps | | Meadow | |
|---|---|---|---|---|---|---|---|---|
| | 1995 | 1996 | 1995 | 1996 | 1995 | 1996 | 1995 | 1996 |
| *Cichorium intybus* L. | 1.0 ± 0.2 | 1.2 ± 0.2 | N.D. | N.D. | 1.5 ± 0.4 | 1,4 ± 0.4 | N.D. | N.D. |
| *Daucus carota* L. | 4.4 ± 1.2 | 3.6 ± 0.5 | 4.8+0.4 | 4.1 ± 0.7 | 4.8 ± 0.8 | 4.0 ± 0.5 | 4.4 ± 0.8 | 4.0 ± 06 |
| *Linaria vulgaris* Mill. | 4.6 ± 0.1 | 4.4 ± 0.2 | N.D. | N.D. | 4.6 ± 0.4 | 4.8 ± 0.4 | N.D. | N.D. |
| *Melilotus albus* Medik | 1.0 ± 0.3 | 1.6 ± 0.4 | N.D. | N.D. | 1.1 ± 0.2 | 1,5 ± 0,2 | N.D. | N.D. |
| *Picris hieracioides* L. | 0.4 ± 0.2 | 0.4 ± 0.1 | N.D. | N.D. | 0.6 ± 0.3 | 0.5 ± 0.2 | N.D. | N.D. |
| *Populus alba* L. | 3.4 ± 0.8 | 3.1 ± 1.2 | N.D. | N.D. | 3.4 ± 0.8 | 3.6 ± 0.5 | N.D. | N.D. |
| *Salix caprea* L. | 3.9 ± 1.1 | 3.4 ± 0.4 | N.D. | N.D. | 3.8 ± 0.4 | 3.9 ± 0.1 | N.D. | N.D. |
| *Trifolium pratense* L. | 3.8 ± 0.6 | 3.7 ± 0.3 | 3.9 ± 0.4 | 4.2 ± 0.6 | 4.0 ± 0.4 | 3.8 ± 0.5 | 3.8 ± 04 | 3.9 ± 0.4 |

Control: Záhorská nížina, Meadow: around 100 meters opposite the sludge bed.
N.D. = not detected.

**Table 7. Clastogenicity of sludge and control solutions (maleic hydrazide and distillated water) evaluated using *Tradescantia* micronuclei test**

| Chemical agents | Dose/Ratio | Type of treatment | MCN/ 100 tetrads (%) |
|---|---|---|---|
| Leach of sludge | 1: 2 | D | 3.93 ± 0.27 * |
| | 1: 2 | I | 3.60 ± 0.43 |
| | 1: 4 | D | 3.67 ± 0.47 |
| | 1: 4 | I | 3.13 ± 0.56 |
| | 1: 8 | D | 2.80 ± 0.27 |
| | 1: 8 | I | 2.67 ± 0.56 |
| Control (+) | 11.2 ųg/ml | D | 5.53 ± 0.82 * |
| | 11.2 ųg/ml | I | 3.67 ± 0.40 * |
| Control (-) | | D | 2.33 ± 0.59 |
| | | I | 2.13 ± 0.49 |

Control (+) = maleic hydrazide, Control (-) = distillated water, D = direct treatment, I =indirect treatment,* significant at 5% level, MCN = micronuclei.

When the results of negative control samples and the results of other samples of the *Tradescantia* micronuclei test (Table 7) and pollen abortivity of the *Tradescantia paludosa* (Table 8) are compared, they showed a significantly higher frequency of micronuclei/ abortiveness of pollen grains after a direct treatment of the maximum concentration (mud extract without dilution) and two and more times higher frequency after a treatment of control mutagen (both direct and indirect exposition). If we compare the indication statement of mitosis and meiosis, we will see comparative results. It confirms that using pollen grains (microspores) by pollen mother cells is also suitable for indication of genotoxicity.

**Table 8. Frequency (%) of abortiveness pollen grains of *Tradescantia paludosa* after treatment of sludge solutions and controls (maleic hydrazide and distillated water)**

| Chemicals | Dose/Ratio | Type of treatment | Abortivity of pollen grains (%) |
|---|---|---|---|
| Leach of sludge | 1: 2 | D | 4.60 ± 0.80 |
| Leach of sludge | 1: 2 | I | 4.20 ± 0.62 |
| Leach of sludge | 1: 4 | D | 4.33 ± 0.85 |
| Leach of sludge | 1: 4 | I | 3.93 ± 0.87 |
| Leach of sludge | 1: 8 | D | 3.67 ± 1.06 |
| Leach of sludge | 1: 8 | I | 3.53 ± 0.88 |
| Control (+) | 11.2 ųg/ml | D | 6.67 ± 1.27 * |
| Control (+) | 11.2 ųg/ml | I | 5.33 ± 0.68 * |
| Control (-) | | D | 3.67 ± 0.27 |
| Control (-) | | I | 3.60 ± 0.30 |

Control (+) = maleic hydrazide, Control (-) = distillated water, D = direct treatment, I = indirect treatment,* significant at 5% level.

## Conclusion

Baldantoni et al. (2014) showed that some plants are good accumulators of pollutants. They found in leaves of *Quercus ilex* L. and *Olea europaea* L. the impact of the Cement factory emissions. Pollutant concentrations was substantial in the area closer to clinker production and storage with the highest Pb, Ni, V, Cr, Fe, indeno(1,2,3–c,d)pyrene, benzo(*g,h,i*)perylene and benzo(*a*)anthracene leaf concentrations. Alfeihm and Wilkström (1984), according to their results from the Ames-test came to the conclusion that the responsibility for the mutagenicity found in the vicinity of Aluminium factory in Norway is at emissions of PAH. However, our observation is that bioindicators, which statistically significantly detect the presence of genotoxic factors "*in situ*," after growth in a contaminated environment through several vegetation periods, are in some cases less sensitive (Mičieta 1992b, 1993). A long-term monitoring is showing an enormous ability of Nature to recover even from such huge pollution like in Žiar nad Hronom, the most polluted part of Slovakia in the past, where aluminum dust is protecting a green life from proliferating in the surrounding area for decades. The frequency of abnormal pollen grains in the control area of Záhorie lowlands (west Slovakia) was found within the norm (up to 5%) during almost eighty years from 1926.

In the all cases of this study we also had to deal with a possible adaptive response, especially regarding the chronic exposure of selected species in the polluted area near the Smelting Foundry (Mičieta and Murín 1997c). In the locality of Banská Štiavnica pollen grains of both species *P. sylvestris* and *P. nigra* tolerated higher concentrations of Pb in the cultivating media during germination (Holub and Ostrolucká 1993). Nearly the same results were found in twenty populations of *Picea abies* (L.) Karst. under the different degrees of industrial pollution in Slovenia (Bavcon et al. 1992). The same species was in focus of a report of Ostrolucká and Fleischer (2003) from a central part of the High Tatra Mts. that, although claimed to be a climatic spa, are affected by long-rank pollution from the heavy industrial areas around the Polish coal-basin (Ciezsyn, Katowice) approx. 200 kilometers to the North-West. Adaptation strategies of Scots and black pine populations subjected to heavy metal pollution were confirmed also by Chudzińska et al. (2014a, b).

While acute treatment by nickel has a significant impact to the pollen germination and pollen tube length of plant species like *Arabis alpina* (Brassicaceae), the ability of such sensitive plant parts like pollen grains to show a significant tolerance to this type of pollution was confirmed by other studies too (Holub and Ostrolucká 1993). Moreover the frequency of their abnormalities as compared to the samples from unpolluted areas could become insignificant (Abramov and Schevschenko 1992).

In summary, we can conclude that particularly arborescent species like *P. sylvestris* and *P. nigra* are suitable as bioindicator species in Slovakia. The results show that the impact of pollution is visible in pollen grain morphology and needle-base meristematic cells. The possible adaptation and/or tolerance of the exposed populations in all industrial areas tested in our study regardless of species of trees or other tested plants cannot be therefore neglected.

*Chapter 4*

# THE BIOINDICATION OF POLLUTION IN URBAN-INDUSTRIAL COMPLEXES

## 4.1. INTRODUCTION

Pollution in urban areas is caused by industry concentrated there and by heavy traffic too. As the content of this pollution in air cannot be clearly defined, wild plant species are optimal objects of testing the bioindication of these mixed pollutants.

In the detection of ecogenotoxicity "*in situ*", besides of the implantation of standardized indicator models like *Tradescantia* and/or *Zea* (cf. Mišík et al. 2007), selected species of the local flora, especially shrub species have been used as bioindicators of phytotoxicity, genotoxicity and mutagenicity in the urban environment (Hool and Cairns 1995, Houghton 1995, Sparling 1995, Mičieta and Murín 1996a, 1997b, 1998a, 2004, Fleck et al. 2016b). For example leaves of *Quercus ilex* L. showed high heavy metal and PAH concentrations also in the urban site, in relation to the vehicular traffic emissions and depositions (De Nicola et al. 2015). Calzoni et al. 2007 used *Rosa rugosa* plants in rural area near Faenza (Italy) by placing homogeneous individuals of this perennial shrub in different sites throughout the area. Soil and leaves of *R. rugosa* were analyzed for their heavy metal content. The total heavy metal pattern of leaves closely paralleled the pattern registered in soil, with the highest content (both in total and assimilable forms) at the site in the urban area, which is exposed to heavy traffic. Makineci et al. (2008) tested survived herbaceous plant species compacted skid road in a fir (*Abies bornmulleriana* Mattf.) forest. The acute damage of plant indicators, after a short-term exposition of high level of pollutants causes necrosis, tissue collapse and withering of whole trees. On the other hand a chronic damage is a result of a long-term exposure to low levels of pollutants. We can observe it in organisms with a long life-span such as trees and shrubs suitable for detection of the chronic impact of pollutants. These damages have no evident external signs. They are manifested as changes of physiological functions. Knowledge of these processes enabled us to use tree and shrub species for long-term monitoring. They can be counted among other eco-toxicologic tests as stated by monograph in this field by Fargašová (2009). The evaluation of the relationship between these types of urban-industrial activity and pollen abortivity was carried out in several localities in Slovakia – urban-industrial complexes in Bratislava, Ružomberok and Žilina including some other cities.

## 4.2. Materials and Methods

For evaluation of genotoxicity we used the following methods:

- evaluation of the frequency of aberrant and anomalous microspores and tetrade analysis at species convenient for bioindication
- evaluation of clastogenic and mitoclastic effects in meiosis – micronucleus test (for criteria of selection of woody bioindicators species see Mičieta and Murín 1996a)
- evaluation of mitosis damage in the meristematic cells as an endpoints of clastogenic and mitoclastic effects of chemicals in environment (cf. Mičieta and Murín 1997a, 2004).

## 4.2.1. Areas of Testing

### *4.2.1.1. Bratislava*

With woody species we were at first interested in their sensitivity. Our aim was to compare pollen abortivity of several woody species in the urban-industrial complexes of Bratislava, capital city of Slovak Republic. At this locality we evaluated pollen abortivity of *Pinus sylvestris* L. and *Robinia pseudoacacia* L. Their higher sensitivity in comparison to the most of herbaceous species was recorded. In the populations growing at the sites where the highest levels of pollution were expected, the increase of pollen abortivity was recorded. In another part of our work we analyzed the changes of the pollen abortivity of *R. pseudoacacia* in response to the increasing distance from the source of pollution.

Figure 8. Picture from the air of an evaluated site. Spalovňa = Incinerator, Areál Slovnaft = Refinery, Miesto odberu vzoriek = sampling place.

Spatial pattern of pollen abortivity was analyzed in the vicinity of a major traffic corridors near the villages of Červeník and Závod in Záhorská lowland, and in the vicinity of the incinerator in Bratislava. Locality Červeník is situated on a major traffic corridor between Bratislava and Žilina. Village Závod is situated on the other major traffic corridor between Bratislava and Brno. Samples were collected in 50 m intervals if it was possible. Contrary to it, in the increasing distance from incinerating tower of city incinerator Bratislava samples were taken in 75 m intervals.

The first sample was taken in distance 220 m from the incinerating tower. From our previous experiences we knew that the highest pollution was found in the area of Slovnaft refinery, where the city's incinerator is localized too. Therefore, we were interested in evaluating by our method of biomonitoring by means of pollen grains of selected plant species (Figure 8).

#### *4.2.1.2. Ružomberok*

The Ministry of the Environment of the Slovak Republic indicated in its memorandum (No. 112/1993) that Ružomberok is one of 12th most polluted areas in Slovakia; therefore it needs special air protection. Particular studies pointed out a possible effect to the human health (Javorka et al. 2005). This aim, consequently, needs serious screening of impact of emission produced at living cells.

We were conducting two studies here.

1. Two years screening of bioindication effect at the cytogenetic level of plants as well as human cells by regularly sampling from the three sites near to the source of pollution as well as from the control site. The aim was to compare the sensitivity of plant cell bioindication essay with human cell screening method at the example of heavy polluted area around a Paper-mill factory.
2. The comparison of the sensitivity of woody and herbaceous species was carried out also in an urban complex Ružomberok. Several localities there were included sites with a different expected level of pollution such as Rybárpole, Klačno, Crossing, Vojany, Railway station and Town hall.

The material of all diploid tree and shrub species from the unpolluted areas (The Záhorská nížina Lowland – villages Velké Leváre and Moravský Ján) was sampled for control to obtain the data about the spontaneous frequency of microspore abortivity. We compared pollen abortivity of several bioindication species from the exposed localities or vicinity of traffic ways. Sampling, slide preparation and evaluation of ecogenoxicity, was made according to protocol from Mičieta and Murín (1996 a) and indication index from Mičieta and Kunová (2000) as well as Mišík et al. (2007) – see at chapter 2.

## 4.2.2. Sources of Pollution

### *4.2.2.1. Bratislava*

Industrial sites in cities are obvious sources of a pollution. Our laboratory had been focused for a long time on such sites – in Bratislava there are eight with well-known emissions balance of pollutants in comparison to other sources (e.g., heavy traffic) (Table 9).

**Table 9. Emissions balance of pollutants (tons/year)**

| | **1970** | | **1975** | | **1985** | |
|---|---|---|---|---|---|---|
| | SO2 | TZL | SO2 | TZL | SO2 | TZL |
| Slovnaft refinery | 30000 | - | 45000 | - | 27678 | 1125 |
| Zpsl. Energy. záv. – factory | 12000 | 10000 | 11400 | 1000 | 9568 | 636 |
| CHZJD – chemical factory | 3000 | 400 | 2000 | 800 | 764 | 84 |
| Matador – gum production | 1500 | 2400 | 1500 | 2400 | 1640 | 90 |
| ČSD – railways | 3000 | 4000 | 3000 | 4000 | - | - |
| Other sources | 6000 | 3000 | 4000 | 2000 | 1973 | 1580 |
| Domestic users | 8000 | 5000 | 6000 | 3500 | 4000 | 2000 |
| Automobile traffic | 50 | 300 | 100 | 600 | 120 | 800 |

Data from the REZZO Pollutant Release and Pollution Register (Závodský 2007). (SO2 – sulfur dioxide, TZL – solid pollutants).

Figure 9 shows a decrease in the values of the basic pollutants for the period between 2000 – 2013 in the district of Bratislava.

City centres are the first to be expected as being affected by air-pollution first by an industrial pollution and/or from heavy traffic. Following are the basic results in mapping Bratislava's emission of the basic pollutants from stationary sources (Table 10) and from mobile sources (Table 11).

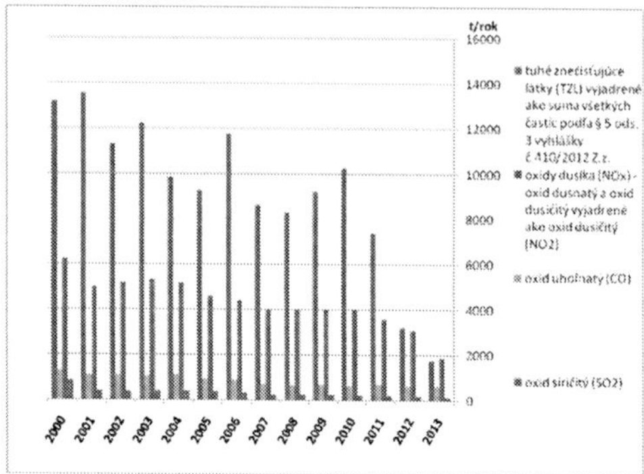


Data in total quantities for tons per year. (TZL= Solid Waste Particles) (Source: NEIS 2014).

Figure 9. Decrease in the values of the basic pollutants for the period 2000-2013 in Bratislava.

## Table 10. Emission of basic pollutants from stationary sources in Bratislava (tons/year)

| | TZL | SO2 | NOx | CO |
|---|---|---|---|---|
| 1990 | 3316 | 33482 | 8603 | 3297 |
| 1995 | 2076 | 27223 | 7985 | 1847 |
| 2000 | 942 | 13240 | 6393 | 1528 |

(TZL – solid pollutants, SO2 – sulfur dioxide, NOx – nitrogen oxides, CO – carbon monoxide) (Závodský 2007).

## Table 11. Emission of basic pollutants from mobile sources in Bratislava (tons/years-1)

| | TZL | SO2 | NOX | CO |
|---|---|---|---|---|
| 1990 | 10764 | 3424 | 56850 | 154199 |
| 1991 | 8855 | 2722 | 47375 | 142135 |
| 1992 | 7978 | 2390 | 43738 | 140621 |
| 1993 | 7064 | 2175 | 42362 | 150676 |
| 1994 | 8544 | 2313 | 43535 | 154804 |
| 1995 | 8755 | 2490 | 45453 | 156743 |
| 1996 | 8940 | 2536 | 45038 | 151133 |
| 1997 | 9142 | 2554 | 44914 | 153216 |
| 1998 | 9509 | 2724 | 46210 | 153946 |
| 1999 | 8766 | 1088 | 43225 | 144655 |
| 2000 | 7648 | 869 | 37298 | 121909 |

(TZL – solid pollutants, SO2 – sulfur dioxide, NOx – nitrogen oxides, CO – carbon monoxide) (Závodský 2007).

**Table 12. Frequency of vehicles in both directions in 24h**

| rok | Trnavská - Sabinovská | Šancová - Jelenia | Račianska - Pekná cesta | Bajkalská - Trenčianska |
|---|---|---|---|---|
| 2005 | 41904 | - | - | 48775 |
| 2006 | 46764 | 43757 | 32291 | - |
| 2007 | 49779 | 47360 | 34121 | 54373 |
| 2010 | 50919 | 47213 | - | 61544 |
| 2011 | 49675 | 48150 | 36061 | 59667 |
| 2012 | 51044 | 47990 | 35535 | 62173 |
| 2013 | 48858 | 45007 | 35461 | 55168 |
| 2014 | 50284 | 49309 | - | 58844 |

Source: Útvar dopravného inžinierstva hlavného mesta SR Bratislava, 2014.

As it is clear that mobile sources are causing significantly higher pollution, we were interested in the frequency of vehicles in the centre within 24h (Table 12).

Pollution in a city centre is also affected by a long-time tendency in weather conditions – rain and temperature (see Figure 10 and 11).

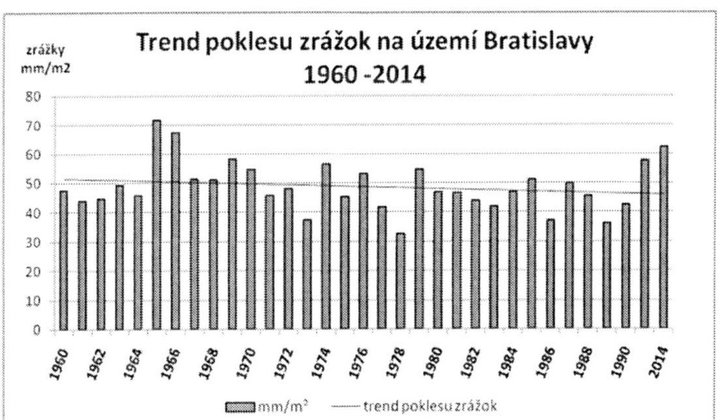


Source: SHMÚ = Slovak Hydrometeorological Institute.

Figure 10. Long-term decrease in rainy precipitation. Measuring station at Bratislava-airport

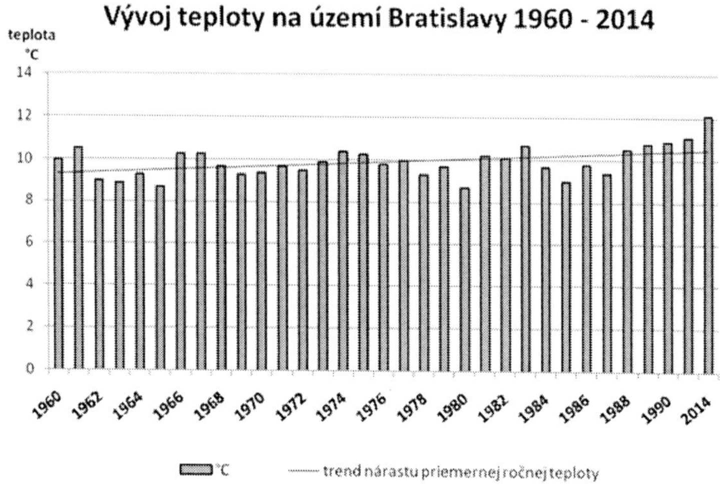


Source: SHMÚ = Slovak Hydrometeorological Institute.

Figure 11. Temperature in Bratislava. Measuring station at Bratislava-airport.

### *4.1.2.2. Ružomberok*

The SCP a. s. (Slovenské celulózky a papierne = Paper-mill factory) in Ružomberok city is the biggest industrial source of the air pollution in the Ružomberok area. It emits into the air not only the basic air pollutants, but in addition organosulphur compounds, inorganic compounds of chlorine, and odour compounds such as $H_2S$. Table 13 shows that SCP is one of three biggest sources of air pollution in the affected area.

**Table 13. Air pollution, quantity of emissions in district of Ružomberok in 1998 (in tons/year). (From Slovak Hydrometeorological Institute screening)**

| | **Solid substances** | **SO2** | **NOx** | **CO** | **Volatile organic substances** |
|---|---|---|---|---|---|
| Stationary sources | 1 189.0 | 2 495.0 | 1 235.0 | 3 304.0 | 536 |
| Mobile sources | 46 | n.d. | n.d. | n.d. | 497 |
| Paper-mill plant | 588.5 | 1 414.1 | 721.5 | 1 729.7 | n.d. |

n.d. = not detected/not determined.

Combined with other stationary sources and road traffic (road E-18) the air pollution often reaches smog level, especially in autumn and winter (Gregušková and Mičieta 2013). Emission measures also show values of $SO_2$ and $NO_x$ over the maximum recommended one-day limit. For this reason we focused our evaluation of ecotoxicity using wild plant species in this area.

### 4.2.3. Collection of Samples and Evaluation

#### *4.2.3.1. Bratislava*

Our preliminary reports uncovered that the trees in general are sensitive to the genotoxic damage. In this study we were focused more deeply on the genotoxic damage of economically important tress, especially on the distribution pattern of pollen abortivity along gradient of pollution and within population at one site and on genotoxic impact of various types of pollution. In the first part we compared the sensitivity of herbaceous and woody species (*P. sylvestris*, *Robinia pseudoacacia* L., *Corylus avellana* L.) from the impacted locality. In a further effort to find the distribution pattern of damage, we tried to assess the changes of pollen abortivity with increasing distance from the source of pollution. To evaluate which types of industrial pollution are connected with increase of pollen abortivity we were testing a pollen abortivity in species *R. pseudoaccacia*, and *P. sylvestris*, in several localities with different types of pollution. Finally we had to analyze the variability of the response at the population level in *P. sylvestris* and *R. pseudoaccacia*.

When we could not find the sufficient amounts of the recommended species at this locality, we may use other species if their pollen abortion at the control site does not surpass 5% and after their precise taxonomical determination. To comply with our method we could also successfully use park trees and shrubs for the genotoxicity indication in the urban sites.

For testing of the impact of motorways in capitol city Bratislava during the given period, we identified from 33 species 23,694 pollen grains in the atmosphere of Bratislava in 2002. In 2003, the total amount of pollen grains was 31,364 (29 species). In 2004 monitoring 34,919 pollen grains were recorded from 32 species, in 2005 the amount represented 40,259 pollen grains from 30 taxa and in 2006 we monitored up to 44,165 pollen grains (31 species), which represented almost twice of the total amount of pollen grains evaluated in 2002.

Dominant in concentrations of pollen grains in air during the monitored years 2002-2006 were taxa: *Ambrosia*, *Artemisia*, *Betula*, *Fraxinus*, and taxa families: *Pinaceae* and *Poaceae*, *Populus*, *Salix*, *Taxus* and *Urtica*.

Most of the pollen grains in the air were recorded in April, when *Betula* was the highest share of the number in all monitored years (56% in 2002, 65% in 2003, 75% in 2005, 39% in 2005 and 67% in 2006).

Collection of samples and their evaluation followed our standard method described previously.

#### *4.2.3.2. Ružomberok*

From the 12 plant species analysed, significant difference from the control samples were found in frequency and occurrence of abortive and anomalous microspores and pollen grains among 9 species in year 1999 and 8 species in year 2000. They were: *Pinus sylvestris* L., *Robinia pseudoacacia* L., *Corylus avellana* L., *Linaria vulgaris* Mill., *Melilotus albus* Medik., *Trifolium pratense* L., *Calystegia sepium* R.Br., *Chelidonium majus* L., *Cichorium intybus* L. In the *Pinus sylvestris* L. samples we observed an increased amount of anomalous (unreduced, 2n, 4n) pollen grains, which represent mutagenity at the genome level. Flowers of the selected indication species from three localities in Ružomberok were collected monthly during their blooming period. Up to 3,000 microspores were evaluated from each monthly sampling of each locality.

As adding test the frequency of aborted/altered pollen grains and chromosomal aberrations in peripheral lymphocytes in blood samples of children living in neighbourhoods of Paper-mill factory were tested. Spontaneous level of chromosomal aberrations in peripheral lymphocytes of children in control population in Slovak Republic and Czech Republic is up to 2% of aberrant cells. Persons who have values of 2-4% of aberrant cells are classified into group with increased exposition to genotoxic substances and persons with values over 4% are considered as group with high exposition to genotoxic substances. From this point of view children from two suburbs (A = suburb Roveň; B = suburb Kľačno) belong to group with increased and children from Lisková (C) to group with high exposition to genotoxic substances. The study was performed with three children populations living in different suburbs (A, B, C) of Ružomberok city situated near of Paper-mill factory and in control population. Each group have contained 25 children in age 5-6 years. Group A represents population from suburb Roveň (South of source of pollution), group B suburb Kľačno (North of source of pollution) and group C village Lisková (East of source of pollution). The control group is represented by children living in village Liptovská Lúžna, which is located 20 km from Paper-mill factory.

Prior to the study, children were examined and only clinically healthy one were included in the study. Children undergoing medical treatment, vaccination or radiography up to 3 month before sampling were excluded.

Heparinized blood samples were obtained from each person. 0.5 ml of whole blood was incubated with 5 ml of RPMI medium (SEVAC Praha) supplemented with 20% inactivated fetal calf serum, antibiotics and glutamine. Cell division was stimulated by 1% phytohemaglutinin (Abbot, HA15). The cultures were incubated for 48 hours at 37°C. Two hours before the end of cultivation, Colcemide (Sigma) of final concentration of $2x10^{-7}$ M was added.The blood cultivation and slide preparation was done according to the method of Natarajan and Obe (1980) modified by Bavorová and Očadlíková (1989). Dry slides were coded and stained with Giemsa-Romanowski for 4 min.

For microscopic analysis slides were cytogenetically analysed under optical microscope with 1,000x magnification. A total of 100 well-spread metaphases containing 46 centromeres were examined per donor. All the basic chromosomal abnormalities, chromatid and chromosome gaps, breaks and exchanges were recorded according to ISCN (1995). To avoid any controversies in the classification, gaps were not included among aberrant cells. The chromosomal damage was expressed as % of aberrant cells and number of breaks per cell.

## 4.3. Results and Discussion

### 4.3.1. Impact of Motorways Around of Capitol City of Bratislava

Figure 12 presents the distribution pattern of the pollen abortivity in *R. pseudoacacia* and its relationship to the increasing distance from the road communication. Two localities included, Červeník and Závod, are situated in major traffic corridors. However, in locality Závod no significant increase of pollen abortivity in comparison to control was recorded. In this locality the samples were collected in gradient Eastward from the traffic corridor, so possibly microclimatic factors are responsible for this result. In locality Červeník the highest relative increase of pollen abortivity was recorded in the distance 200-250 m from the source of pollution. In first 100 m only slight increase of pollen abortivity was recorded followed by relatively steep increase from 100 to 250 m. The fact that road traffic pollution might be one of the factors causing DNA damage in plants was emphasized already by Sriussadaporn et al. (2003). Reponen et al. (2003) analyzed changes of the concentration of the airborne ultrafine particular pollution originating from traffics, which contains several components that could cause DNA damage. Authors observed highest concentration at 80 m from the road and from this distance firstly steep and then moderate decrease of concentration. They presume that at the distance 1600 m background concentration of airborne particles is not achieved yet. In this case as well in immediate distance from the road communication only slight increase of airborne particles were recorded. The mechanism of the induction of the heritable mutation by airborne particles in urban air is discussed by Samet et al. (2004). The fact that the filtration of the airborne particles from the air in the urban industrial area results in lower level of mutations is emphasized by Somers et al. (2004). However, both studies were carried out in animal or human populations, and possibly a lot of the genotoxic substances that are adhered on the surface of the ultrafine particles needs metabolic activation. For evaluating a level of pollution caused by traffic is promising a relatively new technique of bio-magnetic monitoring using roadside plant leaves (Rai 2013).

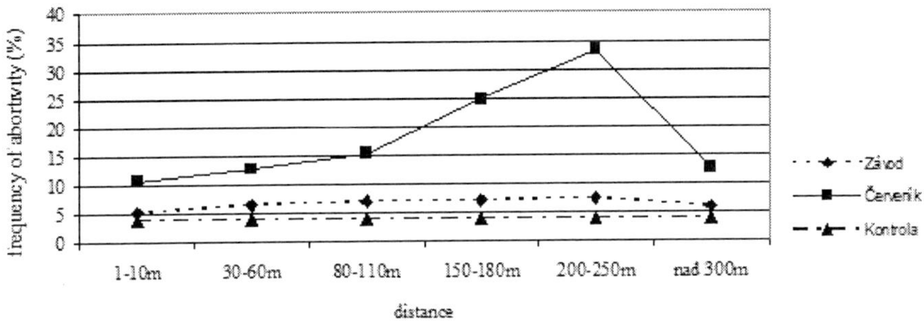


Figure 12. Frequency of pollen abortivity *R. pseudoacacia* in dependence from the distance from the road communication.

## 4.3.2. Impact of Motorways Inside of Capitol City of Bratislava

The lowest concentration of pollen grains was recorded in October and November. As part of the pollen level monitoring in the air at 2-hour intervals during 2004-2006, we recorded the highest concentration of pollen grains in the 12 to 14 p.m. hour time interval. And the lowest concentration from 4 to 6 a.m. hours (Figure 13).

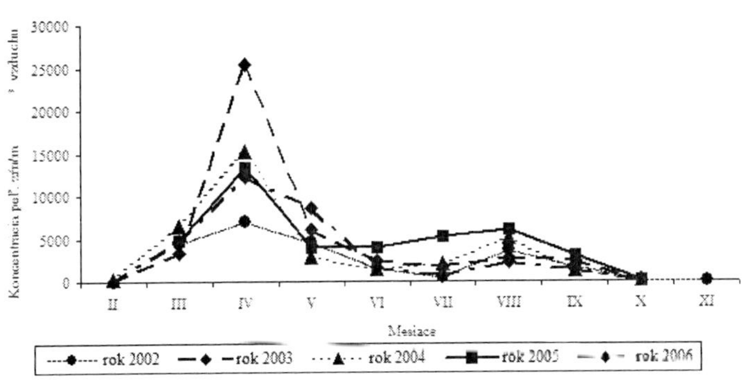


Figure 13. Monthly concentration of pollen grains during pollen season 2002 – 2006.

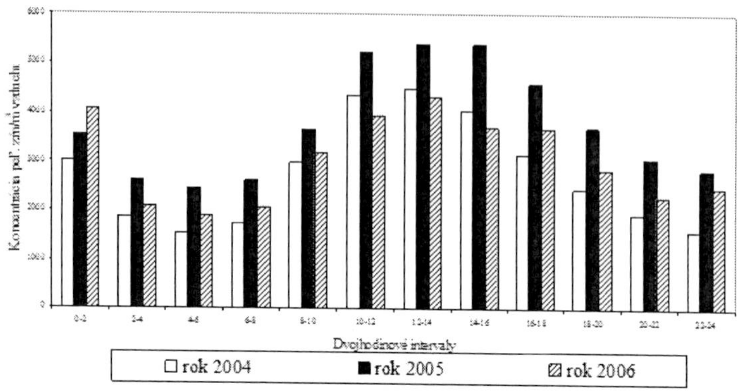


Figure 14. Concentrations of pollen grains per m$^3$ of air at two-hour intervals of pollen seasons 2004, 2005 and 2006.

Figure 14 is showing concentrations of pollen grains per m$^3$ of air at two-hour intervals of pollen seasons 2004, 2005 and 2006.

Our laboratory has been focused on the biomonitoring of the environment of Bratislava city centre since 1996 in yearly intervals. The pollution here had its peak in year 2006 with the same effect at control site. Results are the summary of a whole set of bioindication plant species used (Figure 15).

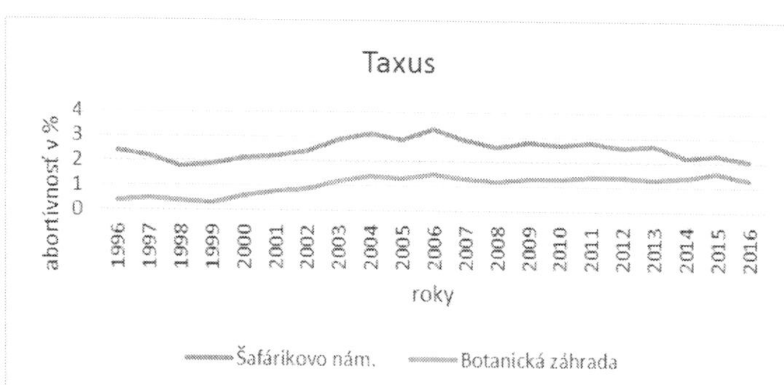


Figure 15. The biomonitoring of the environment of Bratislava city centre since 1996 in yearly intervals till year 2016. (Šafárikovo nám. = square in the centre of the city, Botanická záhrada = Botanic garden in outskirt of the city.)

Present results illustrate an increased ecogenotoxicity in urban ambient, different individual sensitivity and dynamic aspects of the detecting process. Indication index expressly proves of an important increase in ecogenotoxicity in the vicinity of the motorways (Figure 16 and 17).

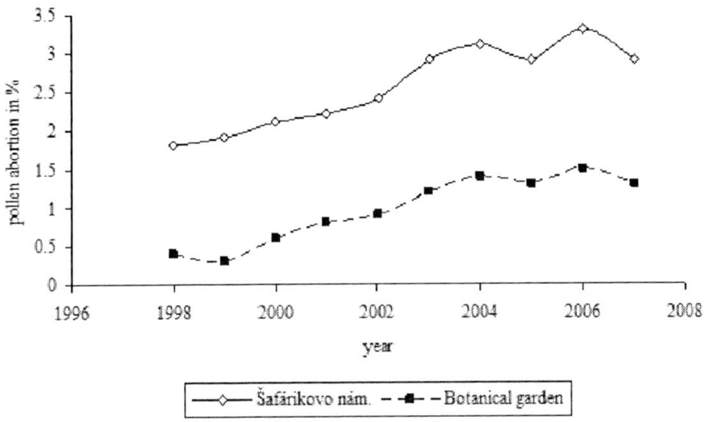


Figure 16. Pollen abortion of *Taxus baccata* in the downtown of Bratislava.

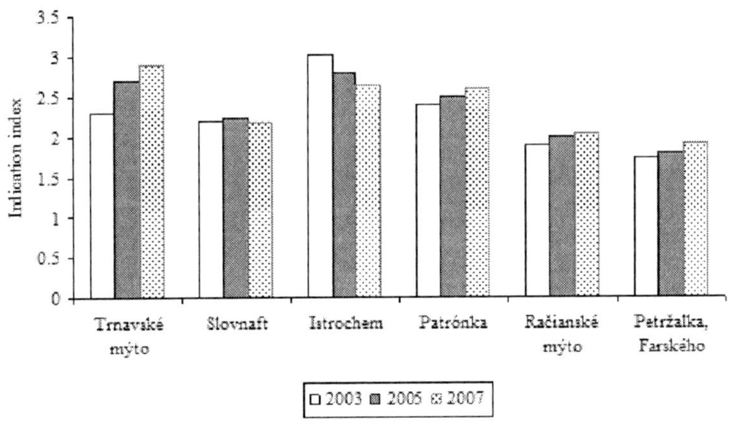


Figure 17. Indication index of selected cross-roads in Bratislava in years 2003, 2005 and 2007.

There are many advantages for using local flora, especially tree and shrub species for the "*in situ*" genotoxicity evaluation (Mičieta and Murín 1996a, Mičieta et al. 2006). Indication of genotoxicity effects at the all structural and functional levels of genetic apparatus of cell:

- Possible indication of pollution or other risk factors in the complex setting of environmental components. It results from the integrative function of plants in the dynamical media.
- Trees and shrubs are organisms with a long life span and accumulate all negative factors from the environment. They are constantly in the environment.
- Possible observation, indication and monitoring of genotoxic /mutagenic effect of risk factors, residuums, or metabolites.
- High sensitivity of plant tests, which are given by the haploid state. All lethal mutations are manifested. We can calculate with mutability of 1,000 locuses.
- Abundance, which allows us to obtain an adequate sample set for competent quantification of the results in the real complex of all ecological factors and types of pollution.
- Indication of genotoxicity effects at the all structural and functional levels of genetic apparatus of cell.

Finally, referring to the obtained data, we may conclude that every linear communication under heavy traffic is a permanent source of emissions causing genotoxic effect in living cells. As our results shows, these may decrease when the traffic is continuous, which is likeable on highways than motorways. Detection and monitoring of traffic pollution is necessary nowadays. Using the native flora for indication is a cheap and approachable method.

### 4.3.3. Results from the Bratislava City Incinerator

#### *4.3.3.1. Data Obtained from Native Flora*

A similar distribution pattern, although not so simple, with an increasing distance from the source of pollution was recorded in locality Bratislava – City incinerator as well (Figure 18). In this case the sampling covered the distance of about 0.7 km and individual samples were collected in intervals 60 – 80 m, first sample was taken at the distance 250 m from the incinerating tower.

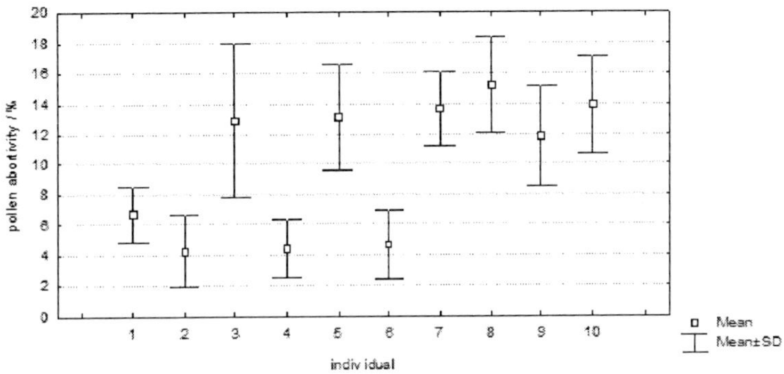


Figure 18. Changes of the pollen abortivity (*R. pseudoacacia*) with increasing distance from the city incinerator in Bratislava. Selected individuals are located in mutual distances 60 – 80 m.

Here, as well as in the close neighborhood of the pollution source no increase of pollen abortivity was recorded and the values of pollen abortivity are similar to the spontaneous pollen abortivity recorded for this species. The interval with no response is followed by the interval where pollen abortivity varies between sampled individuals. In this interval either individuals with pollen abortivity equal to the spontaneous abortivity in *R. pseudoacacia* were recorded, or individuals with 15% pollen abortivity.

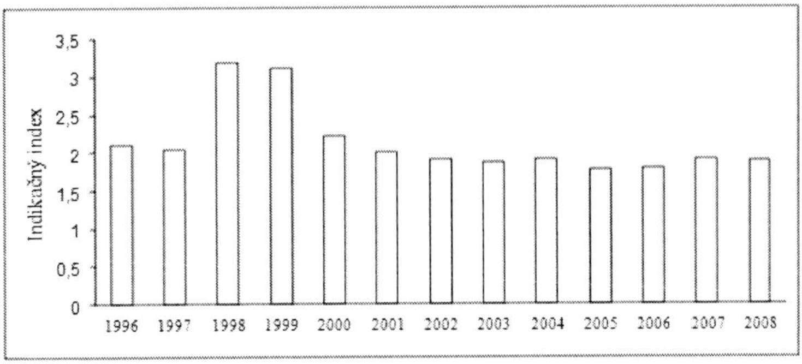


Figure 19. Induction factor of ecogenotoxicity in the vicinity of the Bratislava incinerator in 1996-2008.

Variability of the response in this interval is possibly connected with differential sensitivity of individuals. In another interval starting at distance 0.75 km from the incinerating tower relatively stable percentage of pollen abortivity was recorded (12 – 15%). In this case the results could be biased by the emissions from another nearby source of pollution, the refinery Slovnaft (Jančovičová and Zaliberová 2011). As it is shown in Figure 19 induction factor of ecogenotoxicity is decreasing with following years from the highest peak in years 1998-1999.

The distribution pattern of the pollen abortivity in both cases possibly reflects the pattern of the dispersal of the pollutants from their source. Distribution pattern of "*in situ*" conditions can be affected by variability of the individual sensitivity. High sensitivity of arborescent species to various types of industrial pollution is in concordance with the results of our previous studies (e.g., Mičieta and Murín, 1997b + 1998a, b).

#### *4.3.3.2. Data Obtained from Introduced Tradescantia Test*

Figure 20 is showing that in evaluated years, level of pollution around the city incinerator decreased at a level of distanced sites like Rusovce, which are counted as a norm for a control. The same tendencies are shown at Figure 21 and 22.

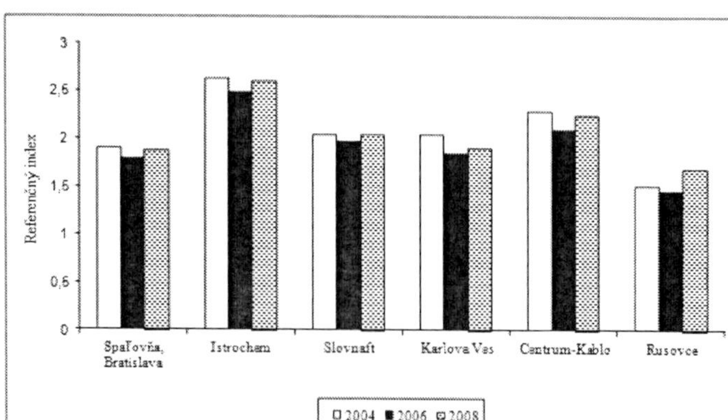


Figure 20. Induction factor of ecogenotoxicity at the monitored localities in Bratislava from 2004 to 2006. Spalovňa = Incinerator, Istrochem = Chemical factory, Slovnaft = Refinery, Karlova Ves = living quarter, Centrum-Kablo = centre of the city, Rusovce = outskirt of the city.

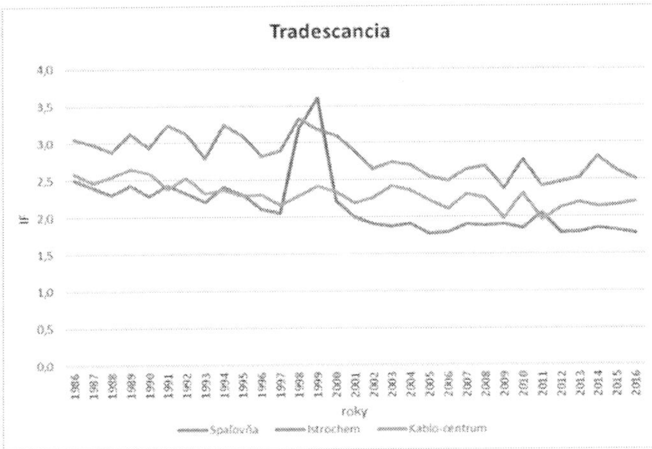


Figure 21. Level of pollution evaluated by Tradescantia at three different sites.

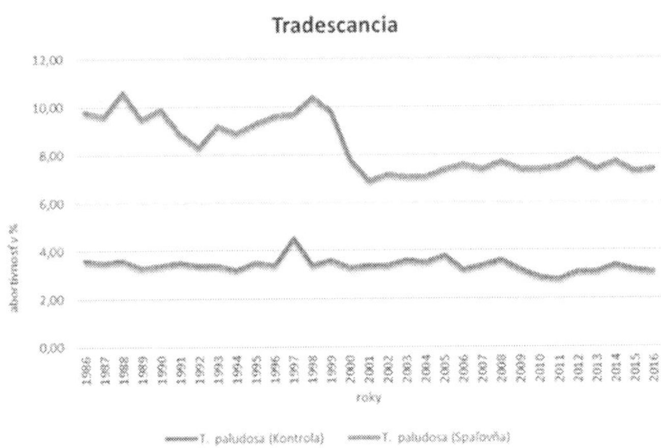


Figure 22. Level of pollution evaluated by Tradescantia at city the incinerator. Control = lower line.

## 4.3.4. Results from Ružomberok

### *4.3.4.1. Ružomberok Paper-Mill Factory*

The whole research was carried out for two consecutive years 1999 and 2000 in the Ružomberok area where pollen abortivity frequency was measured (see Table 14).

Abortion of pollen grains is influenced by the concentration of emissions at the time of meiosis, which is different for each species. Some individuals, populations, and species are able to resist the influence of negative factors, others are more sensitive and could be damaged by a lower concentration of these. The possible adaptation, eventually tolerance of the population, as the result of long lasting chronic or cyclic exposure is possible (Macnair 1987, Mičieta 1992).

It was very interesting to compare the previous results on plant species with standard cytogenetic analysis of peripheral lymphocytes of children in the affected area. The results of these cytogenetic analyses are summarised in Table 15.

**Table 14. Average frequency of pollen abortivity of selected indication species at the localities of Ružomberok in course of 1999 and 2000**

| **LocalitySpecies** | **Roveň A** | | **Klačno B** | | **Lisková C** | | **Control** |
|---|---|---|---|---|---|---|---|
| Year | 1999 | 2000 | 1999 | 2000 | 1999 | 2000 | 2000 |
| *Belis perennis.* | 1.9 ± 0.2 | 1.8 ± 0.2 | 2.2 ± 0.2 | 2.1 ± 0.2 | 2.2 ± 0.2 | 2.2 ± 0.2 | 2.1 ± 0.2 |
| *Corylus avellana.* | 3.2 ± 0.4 | 3.4 ± 0.5 | 3.7 ± 0.5 | 3.8 ± 0.6 | 4.4 ± 0.6 | 4.1 ± 0.6 | 1.4 ± 0.3 |
| *Calystegia sepium* | 1.9 ± 0.2 | 2.0 ± 0.3 | 2.4 ± 0.4 | 2.2 ± 0.3 | 2.7 ± 0.5 | 2.4 ± 0.3 | 1.8 ± 0.2 |
| *Cichorium intybus* | 1.2 ± 0.2 | 1.1 ± 0.2 | 0.9 ± 0.2 | 1.0 ± 0.2 | 1.4 ± 0.2 | 1.2+0.3 | 1.0 ± 0.2 |
| *Daucus carota* | 4.1 ± 0,4 | 3.8 ± 0.5 | 4.2 ± 0.7 | 3.7 ± 0.5 | 4.8 ± 1.0 | 4.5 ± 0.8 | 3.7 ± 0.8 |
| *Chelidonium maj.* | 3.0 ± 0.6 | 2.4 ± 0.4 | 2.2 ± 0.3 | 2.5 ± 0.4 | 5.6 ± 1.2 | 4.4+1.3 | 1.9 ± 0.3 |
| *Linaria vulgaris* | 6.1 ± 2.2 | 5.0 ± 0.5 | 5.3 ± 0.8 | 4.8 ± 0.8 | 9.6 ± 2.3 | 8.4 ± 1.8 | 4.2 ± 0.4 |
| *Melilotus albus* | 3.8 ± 0.6 | 4.2 ± 1.0 | 4.1 ± 0.4 | 4.0 ± 0.6 | 4.8 ± 0.8 | 4.3 ± 0.4 | 1.0 ± 0.2 |
| *Pastinaca sativa* | 2.6 ± 0.4 | 1.8 ± 0.2 | 2.2 ± 0.4 | 2.6 ± 0.3 | 3.2 ± 0.6 | 2.6 ± 0.4 | 2.4 ± 0.4 |
| *Pinus sylvestris* | n.d. | 2.9 ± 0.2 | 2.6 ± 0,2 | 2.8 ± 0.2 | n.d. | 3.2 ± 0.7 | 2.8 ± 0.2 |
| *Robinia pseudoa.* | n.d. | 8.6 ± 2.2 | 9.4 ± 1.6 | 7.1 ± 1.2 | n.d. | 12.4 ± 2.6 | 3.9 ± 0.3 |
| *Trifolium praten.* | 5.1 ± 0.6 | 4.6 ± 0.4 | 4.9 ± 0.7 | 4.7 ± 0.5 | 5.5 ± 0.3 | 5.2 ± 0.4 | 3.6 ± 0.7 |

The spontaneous level of chromosomal aberrations in peripheral lymphocytes of children in the control population in Slovak Republic and Czech Republic is up to 2% of aberrant cells. People who have values of 2-4% of aberrant cells are classified into the group with an increased exposition to genotoxic substances and people with values over 4% are considered as the group with a high exposition to genotoxic substances. From this point of view children from two suburbs (A = suburb Roveň; B

= suburb Kľačno) belong to the group with an increased and children from Lisková (C) to the group with a high exposition to genotoxic substances.

**Table 15. Frequency of chromosomal aberrations in peripheral lymphocytes in blood samples of children living in neighbourhoods of the Paper-mill factory**

| Group | Number of analysed cells | B' | B'' | E' | G | %AB.C | B/C |
|---|---|---|---|---|---|---|---|
| A | 2500 | 41 | 12 | 0 | 17 | 2.20 | 0.020 |
| B | 2500 | 52 | 12 | 1 | 46 | 2.72 | 0.0256 |
| C | 2500 | 82 | 16 | 3 | 61 | 4.16** | 0.0328 |
| Control | 2500 | 31 | 10 | 0 | 31 | 1.88 | 0.0164 |

A = suburb Roveň; B = suburb Kľačno; C = village Lisková; Control = Liptovská Lúžna.

B' chromatid breaks; B'' chromosome breaks; E' chromatid exchanges; G gaps; %AB.C percentage portion of aberrant cells; B/C breaks per cell; **P<0.01.

However, statistical analysis haven't shown significant differences in % of aberrant cells between groups A (t = 1.0746) and B (t = 1.6775) and the control group. Nevertheless, there was a statistically significant difference between the group C (t = 4.0582) and the control group (P<0.01). Considering all obtained data we had come to following conclusions:

1. Both long-term indications, on plant and human cells as well, have shown an impact of polluted environment.
2. Both analyses have shown sensitivity to particular emissions from the Paper-mill factory and additional sources of pollution.
3. Because analyses of plant cells are less elaborative, cheaper and easy to perform as well as to evaluate, it is recommended to use this method for orientative testing that will be then, if an effect founded, proven by more detailed studies on human cells.

### *4.3.4.2. Industrial Complex in Ružomberok City*

Table 16 presents the reference indexes for selected herbaceous and woody species in the urban – industrial complex Ružomberok in years 2004 and 2005. The reference index indicates that the ratio of the pollen abortivity increases in an impacted locality in comparison to the control locality. The highest number of the significantly different frequencies of pollen abortivity was recorded in the locality Rybárpole. In this locality all of the present woody species showed significant results, while from the herbaceous species significantly increased pollen abortivity was recorded just in *Linaria vulgaris* Mill. In two from three analyzed woody species (*C. avellana* and *R. pseudoacacia*) a significant increase of pollen abortivity was recorded in all examined localities. In general, reference indexes in woody species were higher than in herbaceous species except in case of *Melilotus albus* Medik. On the other hand, the variability of the pollen abortivity within single population was higher in this species and therefore a t–test was not significant, what implies the conclusion that *M. albus* is not a suitable indicator. In localities Klačno and Crossing a significant increase of pollen abortivity and relatively high reference indexes were recorded for woody species in comparison to the herbaceous. This could be connected with the fact that microsporogenesis in woody species takes place in a different period then in herbaceous species, so pollen abortivity reflects the pollution levels from a different period.

**Table 16. Reference index of selected herbaceous and woody species, in localities in the urban–industrial complex Ružomberok**

| | Rybárpole | | Klačno | | Crossing | | Vojany | Railway station | | | Town hall | |
|---|---|---|---|---|---|---|---|---|---|---|---|---|
| | 2004 | 2005 | 2004 | 2005 | 2004 | 2005 | 2004 | 2005 | 2004 | 2005 | 2004 | 2005 |
| *Corylus avellana* L. | 2.9 * | 4.2 * | 2.4 * | 4.2 * | | | | | | | | |
| *Pinus sylvestris* L. | | | 0.9 | 0.9 | 3 0* | 1.9 * | 1.0 | 1.0 | 0.7 | 0.7 | | |
| *Robinia pseudoacacia* L. | 3.3 * | 3.6 * | | | | | | | | | 2.7 * | 3.2 * |
| *Bellis perennis* L. | | | 0.7 | 0.8 | 1.0 | 1.1 | 0.7 | 0.9 | 1 | 0.9 | 0.9 | 1.1 |
| *Calystegia sepium* subsp. sepium | 1.2 | 1.5 | | 0.1 | 1.0 | 1.4 | | | 1.1 | 1.2 | | |
| *Chelidonium majus* L. | | | 1.2 | 1.1 | | | 1.1 | 1 | 1 | 1.3 | 1.4 | 1.3 |
| *Cichorium intybus* L. | 0.9 | 1.2 | | | 0.6 | 1.0 | 0.9 | 0.8 | 1.5 | 1.2 | 0.8 | 1.1 |
| *Daucus carota* L. | 1.2 | 1.1 | 0.9 | 1.0 | 1.2 | 1.2 | 0.9 | 1.0 | | | | |
| *Linaria vulgari* Mill.s | 2.5 * | 1.9 * | 1.5 | 1.4 | 1.6 * | 1.6 | 1.4 | 1.4 | 2.1 * | 1.8 * | | |
| *Melilotus albus* Medik. | 4.2 | 4.4 | 4.3 | 4.8 | | | | | 1.0 | 1.6 | | |
| *Pastinaca sativa* L. | 0.5 | 0.8 | 1.1 | 0.8 | 0.8 | 1.0 | | | | | | |
| *Trifolium pratense* L. | 1.6 * | 1.4 * | 1.3 | 1.3 | 1.2 | 1.3 | | | 1.2 | 1.2 | 1.1 | 1.2 |

With * are marked those results, where pollen abortivity in impacted locality was significantly different from control (results were compared using t- test for independent samples).

### 4.3.5. Results from Other Urban Areas

In *R. pseudoacacia* (Figure 23) from localities with medium levels of pollen abortivity the highest intra–population variability was recorded. Three of them are situated in Bratislava and two in other cities. In localities with the highest pollen abortivity the variability tends to be lower at the same time. Relatively high variability of the pollen abortivity in localities in Bratislava can be associated with lower levels of pollution, which some of the individuals can withstand without changing the response, or it may be due to different climatic or other ecologic conditions of this region.

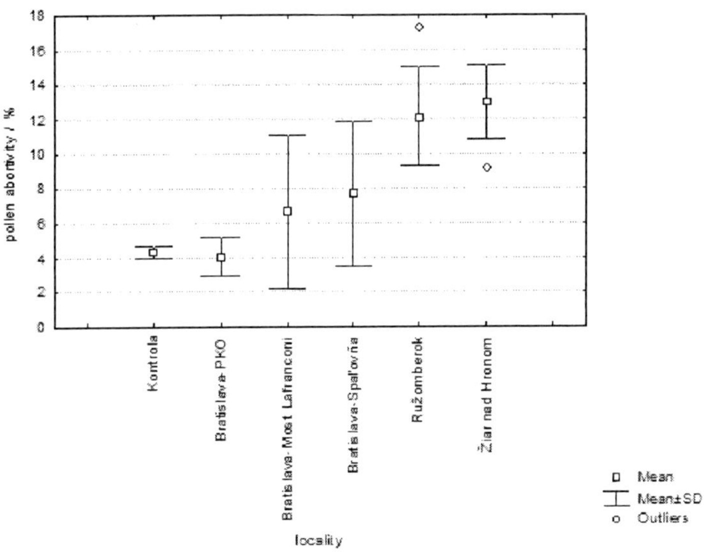


Figure 23. Pollen abortivity in *R. pseudoacacia* in selected localities in Slovakia.

The increase of the variability of the response in relationship to the increased levels of genotoxic pollution was mentioned by Pozolotina (2001). The authoress used a chromosome aberration assay on germinating seeds of *Taraxacum officinale* s. l. from radioactively contaminated zones of Ural Mountains, where the variance of the reply increased in higher levels of contamination. The highest variability of response in populations with medium pollen abortivity was recorded in one of our previous studies (Šípková and Mičieta 2005). Interesting results were obtained from *P.*

*sylvestris* sampled at two localities of Ružomberok and seven other cities (Figure 24).

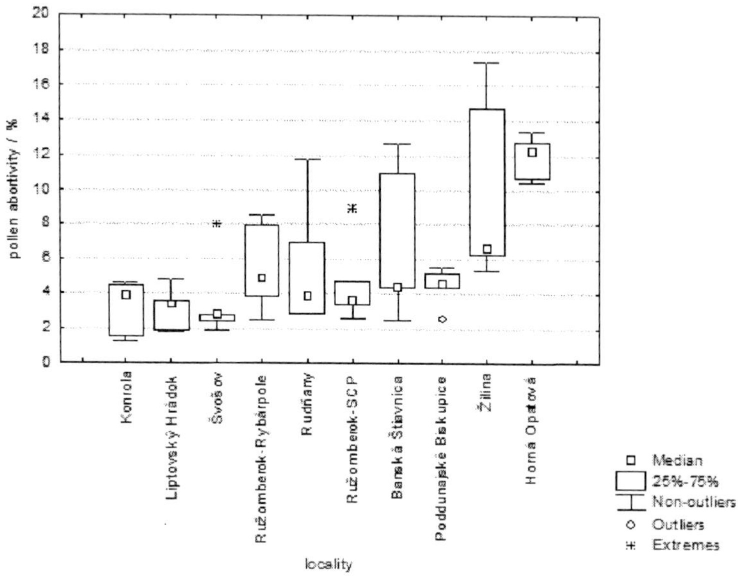


Figure 24. Pollen abortivity in the population of *P. sylvestris*.

# Conclusion

Using plant biomonitoring in cities focused especially to impact of a heavy traffic to the environment with possible consequences for a human health is still rare although these studies are easy to manage with a low cost and universal outcomes (Fleck et al. 2014 + 2016a). Example of these consequences were already shown by Carvalho-Oliveira et al. (2016) using tree bark and pollen abortion rates for assessing air pollution exposure on respiratory mortality rates. Some adaptability of plant species to this kind of pollution was also reported (Demir et al. 2008).

Finally, referring to the obtained data, we may conclude that every linear communication under heavy traffic is a permanent source of emissions causing genotoxic effect in living cells. As our results show, these may decrease when the traffic is continuous, which is more likeable on highways than motorways. Detection and monitoring of traffic pollution is necessary nowadays. The usage of native flora for indication is a cheap and approachable method.

*Chapter 5*

# Wild Plant Species in the Bioindication of Radioactive-Contaminated Sites as Compared with Radio-Resistance of Plants in Laboratory Tests

Our methods (both "in situ" and "in vivo") were tested in three studies:

5.1. Radioactive contamination around nuclear power plant.
5.2. Radioactive contamination from bombardment by low- uranium bombs.
5.3. Laboratory tests of radio-resistance of plants.

## 5.1. Radioactive Contamination Around Nuclear Power Plant

### 5.1.1. Introduction

With the world-wide use of nuclear power plants (NPP) there is often the possibility for radioactive contamination of the adjacent environment. It is not generally known that the first serious failure of nuclear technology with the loss of human lives occurred in NPP Jaslovské Bohunice

(Czechoslovakia, which was a part of the Soviet Block in that time) in January 1976. Until now the worst accident caused by collapse of the NPP in Chernobyl (Ukraine) followed only ten years later (April 1986). Although the catastrophe in Chernobyl raised the public attention, still other nuclear accidents followed with the latest in Japan. In the comparison to these well-known accidents, events like the one in NPP Jaslovské Bohunice were relatively well hidden from public attention. Only after the political changes in 1989 in Czechoslovakia more details appeared.

#### *5.1.1.1. Previous Studies of Impact of the Radioactivity on Plant Cells*

Plants are perspective bioindicators of an environmental impact (Grant 1999) and they can also serve very well for the evaluation of radioactivity in the environment (Minouflet et al. 2005, Murín and Mičieta 2000, Sheppard 2003). Their advantage is in the large number of different plant species and individuals that are around nuclear installations. They are usable as bioindicators immediately after the nuclear accident as well as years later to evaluate its long-term impact. Pollen grains are especially suitable tool for a testing as stressed by other authors too (Vyshenska and Som 2001).

With the world-wide use of NPPs there is often the possibility of radioactive contamination of the adjacent environment. The first well-known accident of this kind appeared in Three Mile Island (March 1979) near Middletown, Pennsylvania (USA); seven years later followed by the nuclear power plant collapse in Chernobyl, Ukraine (April 1986). Although the catastrophe in Chernobyl raised attention of public to nuclear accidents, still other followed – three just in the following month (May 1986 – Hammu-Uentrope, Germany; Sizewel, Great Britain, and Indian Point 2, USA), two in July 1987 (Saint-Laurent-des-Eaux, France and Caceres, Spain), one in December 1987 (Biblis, Germany), one in February 1990 (Creys-Malvile, France), one in February 1991 (Fukui, Japan), one in July 1991 (Kozloduj, Bulgaria), repeated in the same NPP in September 1991, one in September 1999 (Tokaimura, Japan), one in January 2004 (Rovno, Ukraine), one in May 2005 (Great Britain) with the latest in March 2011 (Fukushima, Japan). Despite of all these events only in a few cases the usage of wild plants as bioindicators is known, mostly in Chernobyl.

#### 5.1.1.1.1. Chernobyl in Plant Tests

The catastrophe in the graphite moderated, light-water cooled reactor Chernobyl No. 4 happened in the night from April, 25th to 26th 1986. Soviet officials confirmed in April 1991, that the total death toll was 31 (these people died within 2 months after the accident); 145 people showed signs of radiation-induced illness; and 135,000 people were evacuated within 6 days after the accident.

**Table 17. Test sites and annual dose absorbed by pine crown in mGy**

| Test site | Distance to the Chernobyl NPP | Geographical coordinates | Dγ | Dβ | Dα | Dsum |
|---|---|---|---|---|---|---|
| Refl | ~230 km | 52°46'13,44" N<br>32°39'34,79" E | 0.219 | 0.0058 | 0 | 0.230 |
| VIUA | ~170 km | 52°29'54,42" N<br>31°50'42,31" E | 9.93 | 0.090 | 0 | 10.0 |
| SB | ~190 km | 52°33'52,86" N<br>31°44'21,68" E | 19.2 | 0.175 | 0 | 19.4 |
| Z1 | ~210 km | 53°5'5,78" N<br>31°42'28,29" E | 32.3 | 0.805 | 0 | 33.1 |
| Z2 | ~210 km | 53°5'11,34" N<br>31°42'10,32" E | 38.0 | 0.571 | 0 | 38.6 |
| Kozh | ~45 km | 51° 40' 53,7" N<br>29° 47' 42,1"E | 3.82 | 14.2 | 0.7 | 18.7 |
| Kul | ~20 km | 51° 33' 72,8" N<br>30° 10' 11,0"E | 62.6 | 4.0 | 35.1 | 101.7 |
| Mas | ~12 km | 51° 30' 43,0" N<br>30° 01' 07,2"E | 37.4 | 13.5 | 77.6 | 128.5 |

Dγ – annual dose formed by γ-emitting radionuclides; Dβ - annual dose formed by β-emitting radionuclides; Dα - annual dose formed by α-emitting radionuclides; Dsum – total annual dose formed by α, β, γ-emitting radionuclides. Doses were calculated using the modified dosimetric model described in (Geras'kin et al. 2011).

A land area with a radius of 30 km was closed as an exclusive zone. The total economic consequences of the accident have been estimated up to 25 billion roubels including the relocation of about 200,000 people (Rashad and Hammad 2000). Among the first studies of the impact of the

Chernobyl disaster on the plant species are notable those made by Shevchenko (1989) using *Arabidopsis thaliana*, *Crepis tectorum* and *Tradescantia*. The first and the second species are natural in the vicinity of Chernobyl and thus could be used for the tests of the chronic irradiation. All three species showed the sensitivity to this kind of pollution with a growing response to the higher dose rates. *Tradescantia* test was used also by Cebulska-Wasilewska (1992). *Arabidopsis thaliana* was used later again by Abramov and Schevchenko (1992).

The impact of the Chernobyl radioactive contaminants on the three plant species had been reported by Møller (1998). It was manifested by decreasing of the fitness of tested plants and with a growing level of fluctuating asymmetry and frequency of phenodeviants – these characters are the parameters of stability of an individual development and they depend on the genome stability.

Even the transgenic plants were used as sensitive bioindicators of the nuclear pollution caused by the Chernobyl accident with significant results (Kovalchuk et al. 1998).

Series of the tests in the Chernobyl exclusion zone were done by Sorochinskii (1998) on the protein composition of the anomalous needles of trees like *Picea abies* and *Pinus sylvestris* and coniferous plants generally (Sorochinskii et al. 1998, Sorochinskii 2003). In the co-operation with his colleagues there have been also evaluated pine seeds under the conditions of the chronic irradiation (Isaenkov et al. 1999). Zelena et al. (2005) followed by the tests with *Pinus sylvestris*.

The yield of cytogenetic damage in leaf meristem of agricultural crop plants such as rye and wheat grown in the 10-km zone of Chernobyl NPP was evaluated by Geras'kin et al. (2003) and found to be dependent on the level of the radioactive contamination. Their analysis of experiments with the genetic variability in three sequential generations of rye and wheat revealed increased cytogenetic damage in the plants exposed to chronic irradiation during the 2$^{nd}$ and 3$^{rd}$ years. However, main interest of this laboratory is at *Scots pines* (Geras'kin et al. 2011b). Geras'kin and collaborators showed in several reports also interesting results of a long-term chronic radiation exposure to plant populations at examples of Chernobyl wildlife (Geras'kin et al. 2013, 2015, 2016 and 2018, Garnier-Laplace et al. 2013, Volkova et al. 2018).

Among the latest reports was the comparison of the secondary cell wall structure of tracheids of Scots pine trees (*Pinus sylvestris* L.) prior and after the Chernobyl disaster, showing higher sensitivity of younger trees in alteration of the microfibril angle in $S_2$ and an unusual pattern of the $S_3$ layer (Tulik and Rusin 2005). Even 25 years after the Chernobyl disaster there was still recorded an enhanced absorbed doses in pine crowns in the different distances from the epicentre (see Table 17, Geras'kin et al. 2011, Volkova et al. 2018).

#### 5.1.1.1.2. Other Examples of Practical Use of Wild Plant Tests at Radioactive Sites

Pinawa, Manitoba – The Field Irradiator-Gamma facility in Pinawa, Manitoba, contains jack pine (*Pinus banksiana* Lamb) that has been irradiated. The jack pines retain their seeds from year to year, so that in an irradiated forest, each tree receives a specific dose rate but has seeds that have accumulated a range of total doses. Seed germination and germination rate were examined by Sheppard et al. (1992) on seeds irradiated for up to 5 years. The germination rate was more sensitive and showed deleterious effects at 1.1 $mGy.hr^{-1}$. Hormesis, indicated by a statistically significant increased germination rate, was evident at 0.6 $mGy.hr^{-1}$. Authors stressed that in consideration of the effects on organism other than humans, such as plants, chemical toxicity played also an important role.

Kaiga, India – Two interesting studies were conducted near Kaiga area in the West coast of India where two nuclear power reactors of 220 MW each were commissioned and another two were under construction. The first study was focused towards an investigation of the natural and the artificial fallout radionuclides $^{210}Po$, $^{137}Cs$ and the primordial radionuclide $^{40}K$ in the prominent tree species of this region (*Tectona grandis* L.f. and *Terminalia paniculata* Roth.) and the epiphytic plant species –

*Pterobryopsis tumida* (Hook.) Dix. and *Cymbidium aliofolium* (Lo) Swartz. From a careful analysis of the results, the latest two species were identified as suitable bioindicators to monitor fallout of the radionuclides (Somashekarappa et al. 1996).

Seven years later the same team evaluated the leaves, stem and bark samples from the several plant species collected from the tropical forest of Kaiga and analyzed their $^{226}Ra$, $^{40}K$ and $^{7}Be$ concentrations. The activities of $^{226}Ra$ and $^{40}K$ in the plants were found to vary in the range BDL-13.2 and 12.0-797.3 Bq $kg^{-1}$, respectively. Plants showed significant $^{7}Be$ activity in the leaves, while the stem and the bark of plants showed higher levels of $^{226}Ra$ and $^{40}K$ when compared to the leaves. Soil-to-plant transfer factor for $^{226}Ra$ and $^{40}K$ were found to vary in the range BDL-0.37 and 0.09-5.61, respectively for the different plants. The concentration of $^{226}Ra$ and $^{40}K$ in leaves was dependent on the age of the leaves (Karunakara et al. 2003).

Example of nuclear sites in Russia – Geras'kin et al. (2005 a, b) monitored the genotoxicity of populations of Scots pine in the vicinity of a radioactive waste storage facility in the Leningrad region of Russia (St. Petersburg). During the observation period 1997-2002 pine trees exposed to anthropogenic pollution (with $\gamma$-radiation dose rates from 12.6 $\mu$R/h to 18.8 $\mu$R/h) showed a steady increase of the cytogenetic alterations in the root meristem cells.

Semipalatinsk nuclear test site – For the testing an impact of the chronic radiation exposure on the flora within the Semipalatinsk nuclear test site, Kazakhstan, populations of *Koeleria gracilis* Pers. were used by Geras'kin et al. (2012). It was shown that the frequency of cytogenetic alterations in apical meristem of germinated seeds from the highly contaminated plot significantly exceeded the level observed at other plots with lower levels of radioactive contamination during all three years of the study. A significant excess of chromosome aberrations, typical for radiation exposure, as well as a dependence of the frequency of these types of mutations on dose absorbed by plants were revealed. The results indicate the role radioactive contamination plays in the occurrence of cytogenetic effects. However, no radiation-dependent morphological alterations were detected in the progeny of the exposed populations Geras'kin et al. (2012).

### *5.1.1.3. The Radioactive-Contaminated Sites near NPP Jaslovské Bohunice*

In comparison to known accidents connected with nuclear power plants as mentioned above, the accident in NPP Jaslovské Bohunice was relatively well hidden from the public attention.

Figure 25. Slovakia on the map of Europe.

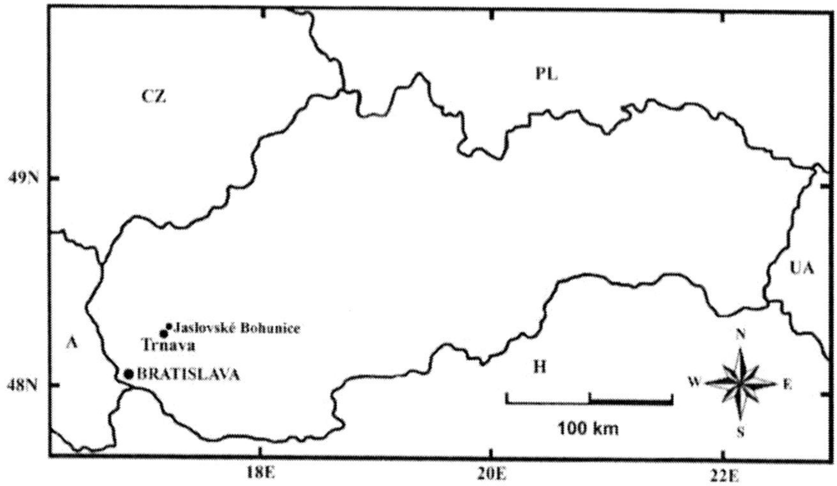


Figure 26. Jaslovské Bohunice on the map of Slovakia.

The first (pilot) NPP in the Slovak Republic, A1, situated in Jaslovské Bohunice site (60 km from Bratislava, the capital city of Slovakia, see Figure 25 and 26) with the capacity of 143 MW, was commissioned in 1972 and was running with interruptions till 1977. In the A1 a KS 150 reactor (HWGCR) with the natural uranium as fuel, D20 as the moderator and gaseous $CO_2$ as the coolant were installed. Refueling was carried out on-line at plant full power as an experimental technology never used in other facilities of this type before and after.

The first serious accident (failure of the closing mechanism of technological channel) happened on January 5$^{th}$, 1976. Fresh fuel assembly (together with the technological plug) was accidentally ejected into the reactor hall. Coolant (carbon dioxide) was flowing out of an open technological channel for a short time until the refueling machine was reconnected with it and the coolant leakage was impromptu stopped (Burclová 1998). This resulted in two deaths from $CO_2$ inhalation. However, the core of the reactor was not damaged and after the reconstruction of the damaged technology and inspection NPP continued in operation in September 1976. Still serious problems were occurring with the integrity of steam generators while the NPP had to be shut down frequently due to failures and subsequent repairs.

The second serious accident occurred in February 22, 1977. During refueling, the insufficiently trans-missive fuel assembly was charged into the core of the reactor. Local overheating of fuel, the channel and heavy water circuit tube caused the loss of the barrier integrity between heavy water moderator and fuel with cooling gas. The cladding and the steam generator tube corrosion under water saturated by carbon dioxide occurred and resulted in the contamination of primary and secondary circuit (Burclová 1998). After this accident the whole A1 reactor was shut down and encapsulated (Štubňa et al. 2002).

Later, four VVER-440s reactors were installed on the same site – in June and January 1981 Unit 1 and 2 (both are the V230 model), followed in February and December 1985 by Unit 3 and 4 (both are the V213 model). The first two units were closed at the end of the year 2008.

The appearance of radioactive-contaminated areas around nuclear power plant Jaslovské Bohunice occurred due to the accident of the A1 block in 1977. During the cleaning up the contaminated material had been accumulated, which was then washed into the near dump canal by heavy raining. While cleaning the canal, the contaminated parts of soil got on the banks of the canal. These places are "blur-like," with 30 cm span and are about 15 cm deep. Their contamination is within 0.067 – 45.5 – 322 kBq/kg $^{137}Cs$.

Information about this accident and following contamination of the neighborhood was kept in secret from 1977 until the collapse of communist regime and even after that it took a few years before occurred an opportunity to reach this site for the close scientific examination. This may happened only thanks to the co-operation with Výskumný ústav jadrových elektrární (VÚJE, Research Institute of Nuclear Power Plants) and Dr. Ondrej Slávik particularly. Initial research was done in cooperation with the Institute of Tumor biology, University of Vienna within the grant Action Austria – Slovakia. Preliminary results were presented at the conferences XIII. Biological days in Hradec Králové, Czech Republic (Murín and Mičicta 1996) and Gmunden am Traunsee, Austria (Murín et al. 1996). Encouraged by vivid response of an academic audience we continued in following years with the more detailed research and the summary of these results is presented in this chapter.

## 5.1.2. Materials and Methods

### *5.1.2.1. Contamination of the Environment*

In both accidents there were large releases of radioactivity. Several nearby villages were affected by tritium contamination of groundwater. Drinking water has been found to contain 65 Bq/litre of tritium, and groundwater near the power plant contains 24 kBq/litre of tritium due to leaking storage tanks. The water beside the tanks contains almost the same radioactivity as the water in the tanks.

After the first accident 4,000 kBq (0.1 Curie) of radioactive iodine $^{131}I$ was released to the environment. In the grass land it was 0.5 kBq/kg $^{137}Cs$ and more than 0.4 kBq/kg iodine $^{131}I$ in the maize land at the distance 11 km from the NPP.

The second accident also induced the releasing of a considerable amount of radioactivity to the vicinity of NPP. In aqueous plants 67,000 Bq/kg of $^{137}Cs$ and 28 kBq/kg of $^{90}Sr$ were observed.

In October 10$^{th}$, 1990, two river bed sediment samples taken 10 km downstream from the reactor were found to contain an average of 0.200 kBq kg$^{-1}$ $^{239}Pu$, $^{240}Pu$ and $^{241}Am$. The samples were taken from the river Dudváh, which flows into the Danube river, the main source of drinking water in Central Europe. During the control in the vicinity of NPP 20 times the normal level of radiation was recorded. The Ministry of Environment of the Slovak Republic called for an immediate action (Gulis 1991).

Hölgye and Filgas (1993) determined the activity concentrations of $^{238}Pu$, $^{239}Pu$, $^{240}Pu$, $^{241}Am$, $^{242}Cm$ and $^{244}Cm$ during 1988 and 1989 in airborne effluents from V-1 reactor. It was found that these radionuclides were present in airborne effluents in the range from 1.0 to 50.0 $\mu Bq.m^{-3}$. In 1993 it was from 3.0 to 26.0 $\mu Bq.m^{-3}$ (Hölgye et al. 1995).

Rather high concentrations of $^{137}Cs$ and $^{90}Sr$ (2150 Bq/dm$^3$ and 215 Bq/dm$^3$, respectively) were found in the bottom sediments of a former channel of the re-engineered river body at a distance of about 250 m from a village Siladice with water-supply wells (Štamberg et al. 1997).

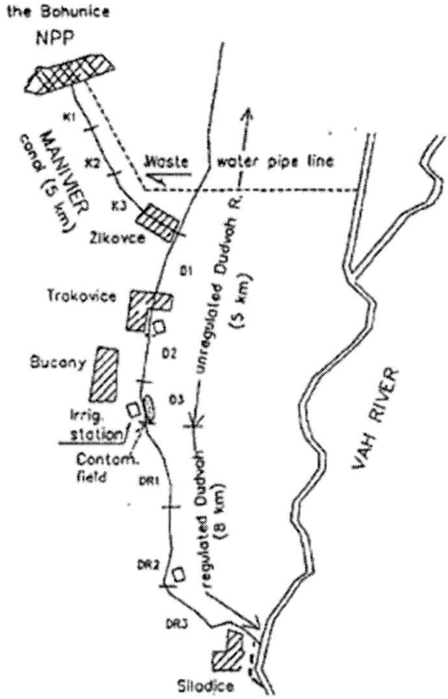


Figure 27. Scheme of the water system taking out the waste water from the Bohunice NPP to the Vah river (Slávik and Morávek 2000).

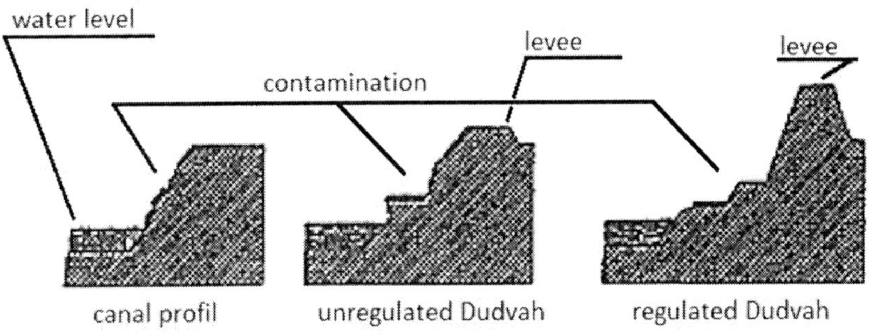


Figure 28. Width profiles and the location of contamination in typical sections of flow (Slávik and Morávek 2000).

The biggest study of this area was done in the cooperation between the International Atomic Energy Agency (IAEA) and VÚJE (Výskumný ústav jadrových elektrární, Research Institute of Nuclear Power Plants) in Trnava. The 19 km long banks of the Jaslovské Bohunice NPP waste water recipient have been identified as contaminated by $^{137}Cs$ as a result of two accidents on the NPP-A1 unit in 1976 and 1977, respectively. In total, more than 67,000 $m^2$ of river banks (their lower and central parts, see Figures 29 and 30) have been found as being contaminated at levels exceeding 1 Bq $^{137}Cs$/g of soil. Until 1992, NPP waste water have been routed through the 5 km long, concrete paved Manivier canal to the smaller rated Dudváh river ($Q_{average}$ = 1.8 $m^3$/s) which joins the Váh river ($Q_a$ = 150 $m^3$/s) further 13 km downstream, 90 km above river Váh's confluence with the Danube river. As was identified by the authors of the study (Slávik and Morávek 2000) in the period between 1976 and 1978, when both accidents happened, a flood control project on Dudváh river had just been implemented over a length of 8 km upstream of its mouth. In the following 5 km upstream of the river section affected by NPP, the flood control conditions were insufficient and have, hitherto, caused permanent concern of the public. The affected recipient water flows directly through, or passes close by, four villages with approximately 5,000 inhabitants. The contaminated banks have, however, an unrestricted access for about 11,000 residents living in a 3.5 km wide strip alongside the river (see Figure 27 and 28).

Pipíška et al. (2004) evaluated the potential of spontaneously growing vegetation as a tool for decontamination of non-agricultural land in the area of the radioactive wastewater of the NPP Jaslovské Bohunice. They observed that autochthonous grasses (mainly *Dactylis glomerata*), *Achillea millefolium* and members of the *Equisetaceae* family (with small amount of other not taxonomically classified plants covering the contaminated area) accumulated a significant amount of $^{137}Cs$. The average specific radioactivity of the biomass harvested from the highly contaminated area was 1 kBq $kg^{-1}$ in the air-dried biomass. Gamma spectrometric analysis of the soil samples showed that $^{137}Cs$ represented more than 99.5% of the total gamma activity of the contaminated soil. However, their results also show very low cesium extractability from the given soil samples.

### *5.1.2.2. Soil Samples*

In the evaluation of radioactivity "*in situ*" we were focused on $^{137}Cs$ as the most reported source of radionuclides in the relevant literature (Sheppard 2003). The collection of soil samples at site was made thanks to the cooperation with the VÚJE in Jaslovské Bohunice. The dose rate was selectively measured in 1994 for each sample with the usage of the shadowed measurer of dose effect NB 3201 (plastic scintillation detector with a low linear energy transfer from 10 nSV to min. 1 mSV/h, Tesla, Czech Republic) in 30 cm height above the soil in the place of the biggest response of the instrument. The samplings in the following years were made from the same sites that were marked for this purpose. The exposition was selectively measured for each sample and reached values between 46.7 – 57 – 196 – 507 – 571 pGy/sec.

Along a 2 km section of the canal banks, just downstream from the NPP discharge point, frequent and highly intense spots of $^{137}Cs$ contamination have been discovered on the central part of the banks. The spots probably resulted from the past improper clean-up operations when highly contaminated mud discharged from the contaminated NPP-A1 sewage system after the flooding in 1980, was manually excavated with shovels from the paved canal bed and scattered over the central part of the embankment surfaces (Slávik and Morávek 2000). These spots are "blur-shaped" with 30 cm span and are about 15 cm deep. Their contamination of $^{137}Cs$ ranges from 0.067 to 322 kBq kg$^{-1}$.

We collected plants from the six different localities where the level of radioactivity contamination differs significantly. It is important to note that these spots differs not only one from another, but that they are sharply bounded. In case of the locality No.1 only one meter distance caused a significant drop of radioactivity level from 45.50 kBq kg$^{-1}$ to almost nil at the locality No. 1a. In case of the locality No. 3 only 15 centimeters distance caused almost a 16 times less radioactivity level for locality No.

3a. Radioactivity level obtained at the locality No. 2 was the highest among the observed localities (Table 18).

**Table 18. Studied soil samples from the vicinity of NPP Jaslovské Bohunice**

| Locality No. | Site of sampling | Activity 137 Cs kBq.kg-1 |
|---|---|---|
| 1 | Manivier drainage, 30 m from the bridge (village Žlkovce), downstream | 45.50 |
| 1a | Manivier drainage, 1 m apart of the spot above (village Žlkovce), downstream | 0.067 |
| 2 | Manivier drainage, 500m from the bridge (village Žlkovce), downstream | 322.00 |
| 3 | Bučany, 15 m from river Dudváh, downstream | 2.38 |
| 3a | Bučany, 30 m from river Dudváh, downstream | 0.15 |
| 4 | Sediment of old river-bed of river Váh, grassy bottom of river Siladice | 9.50 |

### *5.1.2.3. Species of the Local Flora*

On the monitored localities we could use only some of the recommended bioindicator species (Murín 1995, Mičieta and Murín 1996a, Murín and Mičieta 2000). All species collected on these places, from which there was enough material for the analyses, were included in the tests: *Elymus caninus, Cerinthe minor* L., *Dactylis glomerata* L., *Festuca pratensis* Huds., *Arrhenatherum elatius* L., *Poa pratensis* L., *Lysimachia nummularia* L., *Symphytum officinale* L., *Solidago canadensis* L., *Roegneria canina* L.. In the area of Žlkovce (village close to NPP) it was *Raphanus raphanistrum* L., *Linaria vulgaris* Mill., *Trifolium pratense* L., *Lamium purpureum* L, *Lamium maculatum* L., *Calystegia sepium* L., *Chelidonium majus* L., *Melilotus albus* Med., *Pastinaca sativa* L., *Daucus carota* L., *Elytrigia intermedia* and *Phleum pratense* and a new species, suiting to all the criteria and standardized for this study – *Cerinthe minor* L. The testing of the surface waters of the canal close to site No. 1 also included the species *Potamogeton pusillus* L.

*Chelidonium majus* (to the exposure from homogenous source of $^{137}$Cs with doses of 0.17 and 0.34 $\mu$Gy.h$^{-1}$ for 96 h with the recovery time 96 h) was used as a positive control for the tests of the pollen abortivity and

*Dactylis glomerata* (to the exposure of 0.34 $\mu Gy.h^{-1}$ for 96 h, showing a certain impact only without the recovery time) was used for the cytogenetic tests.

### *5.1.2.4. Sampling and the Evaluation*

The tests of the native species were focused on the pollen abortiveness too. Young and closed flowers and flower buds from a sufficient number of randomly selected individual plants (min. 10 individuals) were collected and fixed in a mixture of ethanol:acetic acid (3:1). After 24 h the fixing solution was replaced by 75% ethanol and the samples then stored until further processed. Generally at least 3,000 pollen grains for every species were evaluated, which means 300 pollen grains per one plant. The pollen grains were stained with 0.05% aniline blue in lactophenol.

These previously standardized tests (Mičieta and Murín 1997) were provided annually within 10 years and results were compared with the method of the cytogenetic analysis of chromosome aberrations in the root tip cells of the primary root meristems of collected species. Root tips were fixed in ethanol:acetic acid (3:1), squashed and stained by the aceto-orcein. On average, 200 ana-telophases (50 in control) per slide were evaluated (one slide from each of the ten collected plants). Chromosome aberrations were evaluated in the ana-telophases based on the occurrence of fragments (F), bridges (B) or both (F+B).

The sampling was provided three times within one week. Control samples for analyses of pollen grains as well as cytogenetic evaluation were collected from minimum five individual plants the same day as in the vicinity of Jaslovské Bohunice, but this time it was 60 kilometers to the south-west bank of the river Danube at the fields near the village Lieskovec – Podunajské Biskupice, suburb of the capital city Bratislava without radioactive contamination and with standard results as the cleanest region of Slovakia in our long-term studies (e.g., Mičieta and Murín 1999).

### *5.1.2.5. Standardized Plant Tests*

Results from the above mentioned wild plant tests were confronted with the results of the standard laboratory plant tests.

**5.1.2.5.1. *Vicia sativa* Test – Parallel Test of Phytotoxicity and Mutagenicity**

The test was realized on all monitored soil samples. Opposite to the standardized method this test was modified in the way that swollen seeds in distilled water have not been cultivated on filtrate paper, but planted in soil samples. The control sample was cultivated in a moistened agroperlit (type of sterile mini-balls) and watered by distilled water.

**5.1.2.5.2. *Vicia faba* Test**

For our study the test was modified in the following way:

- 2 h exposition, evaluation immediately after the treatment, ana-telophases;
- 2 h exposition, evaluation 26 h after the treatment (3 h pre-treatment by 0.05% colchicine, c-metaphases);
- 24 h exposition, evaluation immediately after the treatment, ana-telophases;
- 24 h exposition, evaluation 26 h after the treatment; c-metaphases.

**5.1.2.5.3. *Tradescantia* Test**

*Tradescantia* micronucleus test (Trad MCN) was realized as a standardized method. Segments of standard cultivated plants *Tradescantia paludosa* clone 02 were covered by tested soil and exposed for 12 h at 20 °C (ends of segments were covered by the cotton wool which was damped by pabulum Hoagland 2). Frequency of micronucleuses in tetrads with the frequency of chromosome aberrations in ana-telophases of I. and II. meiotic division were observed.

Cultivation, sampling and evaluation were realized in the following way:

- 12 h exposition, evaluation 20 h after the treatment (e.g., recovery time) for the frequency of micronucleuses per 1,000 tetrads;
- 12 h exposition, evaluation immediately after the treatment of 500 ana-telophases I.;

- 12 h exposition, evaluation 30 h after the treatment (e.g., recovery time) of 500 ana-telophases II.

#### 5.1.2.5.4. *Tradescantia* 4430 SHM System Test

*Tradescantia* color-mutation test (Trad-SHM) was performed by the original protocol where segments were exposed for 24 h at 20 °C similarly as clone 4430 in Trad-MCN test.

Statistical analyses between samples from the control area and the polluted areas were based on the analysis of variance (ANOVA). All evaluations were done under the blind conditions. Further details about the used tests can be found in our previous reports (e.g., Mičieta and Murín 1997) and the others (Minouflet et al. 2005).

### 5.1.3. Results

In our study we were focused on the genotoxic analysis of pollen grains and the cytogenetic tests of the local flora bioindicators "*in situ*" at sites contaminated by radioactivity. As it is shown in Table 19, for the evaluation of pollen abortivity (±SE) we had used nine different species of the local flora. The sensitivity inside of this group to the radioactivity differs 16 times in the abortivity frequency from *Linaria vulgaris* to *Cerinthe minor* from the highest level (4.6 ± 1.1%) to the lowest (0.3 ± 0.1%). However, a significant difference from the control was not found in any of the species. For example, the average frequency of abortivity at the contaminated locality No.2 was 2.74 ± 0.5% while at the control site it was 2.93 ± 0.6%.

Results shown in the Table 20 were from the sampling at the locality No.2 in 1994. We were interested in the situation at this locality six (2000) and ten (2004) years later, respectively. Our set of the indicative species was enlarged for the species *Lamium maculatum* and *Calystegia sepium*. It was because in case of *Cerinthe minor* we were unable to find enough plants at the locality and therefore another two species were added.

**Table 19. Results of the test of pollen abortivity (± SE) in species of local flora from the locality No. 2 in 1994**

| Species | Number of evaluated pollen grains | Frequency of abortivity in % | Control in % |
|---|---|---|---|
| *Raphanus raphanistrum* | 3000 | 3.3 ± 0.8 | 4.4 ± 0.4 |
| *Linaria vulgaris* | 3000 | 4.8 ± 0.8 | 4.6 ± 1.1 |
| *Trifolium pratense* | 3000 | 2.8 ± 0.6 | 3.3 ± 0.4 |
| *Cerinthe minor* | 3000 | 0.3 ± 0.1 | 0.3 ± 0.1 |
| *Lamium purpureum* | 3000 | 3.4 ± 0.6 | 3.0 ± 1.0 |
| *Chelidonium majus* | 3000 | 2.1 ± 0.5 | 2.0 ± 0.3 |
| *Melilotus albus* | 3000 | 1.0 ± 0.3 | 1.0 ± 0.4 |
| *Pastinaca sativa* | 3000 | 2.9 ± 0.4 | 3.4 ± 0.4 |
| *Daucus carota* | 3000 | 4.1 ± 0.8 | 4.4 ± 1.2 |
| Average | | 2.7 ± 0.5 | 2.9 ± 0.6 |

**Table 20. Frequency of abortive pollen grains of species (±SE) of native local flora from the locality No.2**

| Species | Abortive pollen grains (%) | Control (%) | Abortive pollen grains (%) | Control (%) | Abortive pollen grains (%) | Control (%) |
|---|---|---|---|---|---|---|
| | 1994 | 1994 | 2000 | 2000 | 2004 | 2004 |
| *Raphanusraphanistrum* | 3.3 ± 0.8 | 4.4 ± 0.4 | 4.5 ± 0.5 | 4.3 ± 0.6 | 3.8 ± 0.4 | 3.6 ± 0.4 |
| *Linaria vulgaris* | 4.8 ± 0.8 | 4.6 ± 1.1 | 5.0 ± 0.4 | 3.2 ± 0.4 | 4.7 ± 0.5 | 4.2 ± 0.4 |
| *Trifolium pratense* | 2.8 ± 0.6 | 3.3 ± 0.4 | 3.8 ± 0.3 | 3.7 ± 0.7 | 3.8 ± 0.4 | 3.4 ± 0.3 |
| *Cerinthe minor* | 0.3 ± 0.1 | 0.3 ± 0.1 | 0.4 ± 0.2 | 0.4 ± 0.2 | - | - |
| *Lamium purpureum* | 3.4 ± 0.6 | 3.0 ± 1.0 | 3.3 ± 0.4 | 2.9 ± 0.4 | 3.6 ± 0.2 | 3.4 ± 0.4 |
| *Lamium maculatum* | 2.8 ± 0.2 | 2.6 ± 0.2 | 3.1 ± 0.4 | 3.0 ± 0.2 | 3.2 ± 0.4 | 3.1 ± 0.3 |
| *Calystegia sepium* | 1.8 ± 0.2 | 1.6 ± 0.2 | 1.9 ± 0.2 | 1.8 ± 0.2 | 2.1 ± 0.3 | 1.9 ± 0.1 |
| *Chelidonium majus* | 2.1 ± 0.5 | 2.0 ± 0.3 | 2.6 ± 0.4 | 2.2 ± 0.4 | 2.5 ± 0.3 | 2.2 ± 0.3 |
| *Melilotus albus* | 1.0 ± 0.3 | 1.0 ± 0.4 | 1.1 ± 0.3 | 0.9 ± 0.2 | 1.2 ± 0.2 | 1.2 ± 0.2 |
| *Pastinaca sativa* | 2.9 ± 0.4 | 3.4 ± 0.4 | 3.5 ± 0.3 | 3.1 ± 0.4 | 3.8 ± 0.4 | 3.5 ± 0.4 |
| *Daucus carota* | 4.1 ± 0.8 | 4.4 ± 1.2 | 4.4 ± 0.8 | 4.2 ± 1.2 | 4.4 ± 0.3 | 4.6 ± 0.3 |
| Average abortiveness | 2.7 ± 0.5 | 2.8 ± 0.5 | 3.1 ± 0.4 | 2.7 ± 0.4 | 3.0 ± 0.3 | 2.8 ± 0.3 |
| Positive control | Exposure μGy.h-1 | | | Abortive pollen grains (%) | | |
| *Chelidonium majus*x | | 0.17 (96 h) | | | 3.0 ± 0.4 | |
| *Chelidonium majus* | | 0.34 (96 h) | | | 5.4 ± 0.6 | |

*P ≤ 0.05.

**P ≤ 0.01.

x = recovery time 96 h (highest determined pollen abortivity).

**Table 21. The cytogenetic test on species of local flora near the nuclear power plant Jaslovské Bohunice at several localities in 1994**

| Locality | Species | Number of evaluated ana-telophases | Frequency of aberrations in % (±SE) |
|---|---|---|---|
| 1 | *Roegneria canina* | 500 | 0.4 ± 0.4 |
| | *Cerinthe minor* | 500 | 0 |
| | *Dactylis glomerata* | 500 | 0.2 ± 0.2 |
| 1a | *Potamogeton panormitanus* | 250 | 0 |
| | *Elymus caninus* | 500 | 0.2 ± 0.2 |
| | *Cerinthe minor* | 500 | 0.2 ± 0.2 |
| | *Dactylis glomerata* | 500 | 0.2 ± 0.2 |
| 2 | *Elymus caninus* | 500 | 0 |
| | *Dactylis glomerata* | 500 | 0.4 ± 0.4 |
| | *Festuca pratensis* | 500 | 0.2 ± 0.2 |
| | *Arrhenatherum elatius* | 500 | 0.2 ± 0.2 |
| 4 | *Poa pratensis* | 500 | 0.2 ± 0.2 |
| | *Lysimachia nummularia* | 500 | 0 |
| | *Symphytum officinale* | 500 | 0.4 ± 0.4 |
| | *Solidago canadensis* | 500 | 0.2 ± 0.2 |
| Control | *Festuca pratensis* | 500 | 0 |
| | *Arrenatherum elatius* | 500 | 0.2 ± 0.2 |
| | *Poa pratensis* | 500 | 0.2 ± 0.2 |
| | *Lysimachia nummularia* | 500 | 0 |
| | *Symphytum officinale* | 500 | 0.4 ± 0.4 |
| | *Solidago canadensis* | 500 | 0.2 ± 0.2 |

As it is shown in the Table 20, no significant response to the radioactive contamination in the bioindicator species tested for pollen abortiveness at the locality No. 2 was found in 1994. This result was confirmed within the following six and ten years of testing. Again, the most sensitive to the radioactive contamination appeared to be *Linaria vulgaris* and the lowest impact was found again at samples from *Cerinthe minor* with more than 16 times difference. However, these results do not differ significantly from the control samples, although it was at the locality with the highest radioactive contamination. An important role here plays also an individual variability, where the level of pollen abortivity up to 5% may be generally counted as spontaneous due to the climatic condition "*in situ*" (Murín and Mičieta 2000b). The average frequency of the abortivity at the contaminated locality No.2 was 2.66 ± 0.5% for year 1994, 3.05 ±

0.4 % for year 2000 and 3.01 ± 0.3% for year 2004, while at the control site it was 2.78 ± 0.5%, 2.70 ± 0.4% and 2.83 ± 0.3% respectively.

In this case we have tested also the positive control with the species *Chelidonium majus* showing the dose dependent response after 96 h of recovery time. In case of the cytogenetic analyses of the primary root meristems we have selected ten different plant species from four localities according to the standardized method (Mičieta and Murín 1997).

The evaluation of all collected species showed that results were insignificant in 1994 (Table 21) as frequency of aberrations never exceeded spontaneous level characteristic for the control samples. The same situation was for all four localities where even the level of contamination by the radioactivity differs between them from the lowest (0.067 kBq $kg^{-1}$) to the highest (322.00 kBq $kg^{-1}$).

**Table 22. The cytogenetic test of local flora species from the locality No. 2 in the vicinity of the NPP Jaslovské Bohunice in 2004**

| Species | Number of evaluated ana-telophases | Frequency of aberrations in (%) ±SE |
|---|---|---|
| *Roegneria canina* | 500 | 0.2 ± 0.2 |
| *Dactylis glomerata* | 500 | 0.4 ± 0.4 |
| *Festuca pratensis* | 500 | 0.2 ± 0.2 |
| *Arhenatherum elatius* | 500 | 0.2 ± 0.2 |
| *Elytrigia intermedia* | 500 | 0.2 ± 0.2 |
| *Phleum pratense* | 400 | 0.5 ± 0.2 |
| Negative control | | |
| *Roegneria canina* | 500 | 0.2 ± 0.2 |
| *Dactylis glomerata* | 500 | 0.2 ± 0.2 |
| *Festuca pratensis* | 500 | 0.2 ± 0.2 |
| *Arhenatherum elatius* | 500 | 0.2 ± 0.2 |
| *Elytrigia intermedia* | 500 | 0.2 ± 0.2 |
| *Phleum pratense* | 500 | 0.4 ± 0.4 |
| Positive control | | |
| *Dactylis glomerata** | 300 | 2.3 ± 0.3 |

* without recovery time (with recovery time was not increased clastogenicity).

From the localities shown in the Table 18 we had chosen the one with the highest radioactive contamination (No. 2) and evaluated it again ten years later, in 2004. For this we had enlarged the spectra of bioindicator species from four to six with three new ones (*Roegneria canina, Elytrigia intermedia* and *Phleum pratense*). However, in this locality the results were again insignificant while the positive control on *Dactylis glomerata* showed a significantly higher response to the 0.34 $\mu Gy.h^{-1}$ (Table 22).

All "*in vivo*" tests used were showing no significant impact of the radioactivity from soil samples on their genotoxic end-points (see Table 23–26).

**Table 23. Results of "*in vivo*" tests**

| Sample | IC in % | No of ana-telophases | Frequency of aberrations in % |
|---|---|---|---|
| 1 | 0 | 500 | 0.4 ± 0.4 |
| 1a | 0 | 500 | 0.2 ± 0.2 |
| 2 | 0 | 500 | 0 |
| 3 | 0 | 500 | 0.2 ± 0.2 |
| 3 a | 0 | 500 | 0.4 ± 0.4 |
| 4 | 0 | 500 | 0.2 |

**Table 24. Results of "*in vivo*" tests**

| Sample No. | Activity 137Cs pGy/sec | Exposition in hrs | No. of evaluated ana-telophases*c-metaphases | Frequency of aberrations in % |
|---|---|---|---|---|
| 2 | 571 | 2/0 | 300 | 0.3 |
| 2 | 571 | 2/26 | *300 | 0.3 |
| 2 | 571 | 24/0 | 300 | 0.6 |
| 2 | 571 | 24/26 | *300 | 0.3 |

**Table 25. Results of "*in vivo*" tests**

| Sample No. | Activity 137Cs pGy/sec | Exposition in hrs/recovery time | No. of evaluated *I. ana-telophases xII. ana-telophases o tetrades | Frequency ofa berrations and MCNo in % |
|---|---|---|---|---|
| 2 | 571 | 12/0 | *500 | 5.6± 0.4+ |
| 2 | 571 | 12/30 | x500 | 3.4± 0.4+ |
| 2 | 571 | 12/30 | o1000 | o3.2± 0.2+ |

**Table 26. Results of "*in vivo*" tests**

| Sample No | Activity 137 Cs pGy/sec | SHM mutations in % 10th day after exposition |
|---|---|---|
| 1 | 507 | 0.26± 0.06 |
| 1a | 46.7 | 0.30± 0.05 |
| 2 | 571 | 0.36± 0.07 |
| 3 | 196 | 0.26± 0.04 |
| 3a | 57 | 0.23± 0.05 |
| 4 | 9.50 | 0.16± 0.04 |
| Positive control malein hydrazide | 0.1M | 11.20± 0.63++ |

In this part of our study we were focused on the genotoxic analysis of the response of the standardized plant species to the soil samples contaminated by radioactivity. As it is shown in Tables 23–26, we can state, that no phytotoxicity or increased frequency of mutations against the control was recorded.

### 5.1.4. Conclusion

Plants are perspective bioindicators of an environmental impact (Grant 1999) and they can also serve for the evaluation of the radioactivity in the environment. Their advantage is that a large number of different species and individuals around nuclear installations are usable as the bioindicators immediately after a nuclear accident as well as years later to evaluate its long-term impact. Most reports of usage of wild plants as bioindicators of these accidents are from Chernobyl.

Among the first studies of the impact of the Chernobyl disaster on the plant species are notably those made by Shevchenko (1989) using *Arabidopsis thaliana*, *Crepis tectorum* and *Tradescantia* (clone 02). The first and the second are present naturally in the vicinity of Chernobyl and thus could be used for the tests of the chronic irradiation. All three species showed sensitivity to this kind of pollution with a growing response to the higher dose rates.

The impact of Chernobyl radioactive contaminants on three plant species (*Robinia pseudacacia*, *Sorbus aucuparia* and *Matricaria peforata*) was reported by Møller (1998). Fluctuating asymmetry and the frequency of phenodeviants showed developmental instability positively related to the level of radiation by $^{137}Cs$ being three to four times larger near Chernobyl as in the control area.

Even transgenic plants were used as sensitive bioindicators of nuclear pollution caused by the Chernobyl accident with a significant results (Kovalchuk et al. 1998).

A series of tests in the Chernobyl exclusion zone were done by Sorochinskii (1998) on the protein composition of the anomalous needles of trees as *Pinus sylvestris* (Kuchma and Finkeldey 2011, Kuchma et al 2011), *Picea abies* and coniferous plants generally (Sorochinskii et al. 1998, Sorochinskii 2003). Pine seeds were also studied under the conditions of the chronic irradiation (Isaenkov et al. 1999).

The yield of the cytogenetic damage in a leaf meristem of agricultural crop plants such as rye and wheat grown in the 10-km zone of Chernobyl NPP was evaluated by Geras'kin et al. (2003) and found to be dependent on the level of the radioactive contamination. Their analysis of experiments with genetic variability in three sequential generations of rye and wheat revealed an increased cytogenetic damage in plants exposed to the chronic irradiation during the 2$^{nd}$ and 3$^{rd}$ year.

Among the latest reports was the comparison of the secondary cell wall structure of tracheids of Scots pine trees (*Pinus sylvestris* L.) prior to and after the Chernobyl disaster showing higher sensitivity of younger trees in an alteration of the microfibril angle in $S_2$ and an unusual pattern of the $S_3$ layer, respectively (Tulik and Rusin 2005).

In contrast to the above mentioned reports, we had found no significant impact on the wild plant species at radioactive-contaminated sites around NPP Jaslovské Bohunice in the Slovak Republic. The explanation is to be found partially in the fact that the culture of high grasses is newly planted on the banks of the canal and the nearby fields are intensively used for farming. We may also note that seemingly high radioactivity found around Jaslovské Bohunice is much lower than in other areas like in Chernobyl (up to 18.5 x $10^5$ kBqm$^{-2}$, see Tulik and Rusin 2005) tested in the same time with the significant results. It is also indicated by other authors that the given soil samples from vicinity of NPP Jaslovské Bohunice show very low cesium extractability (Pipíška et al. 2004). A possible adaptive response to the long-time exposition cannot be neglected too if we know that the chronic effect of radioactivity may lead to the adaptation processes resulting in the radio resistence (Oudalova et. al. 2005). The natural selection of the most adapted genotypes of *Stipa capillata* was observed within almost a half century of the exposure to a low dose rate of the ionizing radiation on the Semipalatinsk nuclear test site in Kazakhstan (Zaka et al. 2002). In our case the first sampling and evaluation occurred (by circumstances not under our control) only 18 years after the accident. The possible adaptive response to the long-time exposition therefore cannot be neglected.

Although the standard laboratory plant tests used by us are known as a sensitive bioindicators of radioactivity (Minouflet et al. 2005), our soil samples did not initialize any significant response in scale of genotoxicity of these plants.

Our results from all these different plants are supported by the special human study of long term incidence of cancer disease in the vicinity of NPP Jaslovské Bohunice. It was conducted in the period from 1980 till 1990 for men and women separately for each municipality in the Trnava district. Neither a higher incidence of cancer nor accelerated dynamics of cancer formation was found. Therefore, the mortality is uniform within the whole Trnava district and no exceptional situation was found there (Letkovičová et al. 1997).

Finally, we may state that there is no phytotoxicity or an increased frequency of mutations in pollen grains or root tip cells of the wild plants at the contaminated sites near NPP Jaslovské Bohunice compared to the control samples. The same can be stated in the case of the standard tests such as *Vicia faba*, *Vicia sativa* and *Tradescantia* used in this study.

# 5.2. Phytotoxicity and Mutagenicity of the Contaminated Locations after NATO Bombarding in Serbia in 1999

## 5.2.1. Introduction

From March 24th, 1999, NATO undertook a total of 36,219 bombings of Serbia for 78 days. The UNEP /UNCHS-(Habitat) Working Group on the Balkans has located four ecological outbreaks of the following damages in Pančevo (refinery), Kragujevci (soil contamination with dioxins), Novi Sad (Danube water contamination) and Bor (copper mine).

This synergistically multiplied the already existing environmental pollution.

Residents not only of Serbia but also Bosnia, Kosovo and Montenegro were confronted with low-level uranium dispersed by American weapons dropped over Bosnia in 1995, and over Yugoslavia during the 1999 Kosovo war. UN representative Pekka Haavisto stated for the world media (Simons 2001): "We found some radiation in the middle of villages where children were playing. We were surprised to find this a year and a half later. People had collected ammunition shards as souvenirs and there were cows grazing in contaminated areas, which means that the contaminated dust can get into the milk."

These revelations concerned not only the affected areas and countries, but also the countries of soldiers serving that time there or in the close range. The 15-countries of European Union have ordered their own inquiry into the possible noxious effects of the uranium-tipped ammunition and any potential link to the recent cancer deaths among Balkan veterans. Tens of thousands of European soldiers who served in the Balkans have undergone quietly conducted medical tests in countries like Belgium, France and Canada. In January, 2001, Italy, the Netherlands, Portugal, Spain, Finland, Norway, Greece and Bulgaria have announced that they will screen all their Balkan veterans.

Belgium officials were the first to report that nine Balkan veterans had fallen ill with the cancer, and five having already died. Two veterans died of the leukemia in the Netherlands, and one in Spain. France said it was treating four veterans for the leukemia. In Italy, 30 veterans contracted serious illnesses, 12 of whom developed cancer. Six of the cancer patients had already died of leukemia (Simons 2001).

As we had an open access to the affected areas in Serbia (Bačka Palanka and Novi Sad) we had started our own study.

## 5.2.2. Materials and Methods

By bombardment, an estimated 135,000 tons of toxic material affected targeted areas, including 30-50 tons of radioactive uranium $^{238}U$, whose alpha radiation is up to 20 times more destructive than gamma-ray bombs (Dalmacija et al. 2000). Therefore, we were focused on the test of the consequences of using these bombs (see Table 27).

To investigate the toxicity and the mutagenity of soil from the craters in Bačka Palanka and Novi Sad, we used a parallel phytotoxicity and mutagenicity test on seeds of *Vicia sativa* L., *Raphanus sativus* L., *Zea mays* L. according to the already established methodology (Murín 1987, Mičieta and Murín 2002b).

**Table 27. Results of the γ-spectrometric analysis of the soil samples in the territory of Novi Sad and surroundings obtained during and after the NATO bombing (Dalmacija et al. 2000)**

| Sample | 238 U | 226 Ra | 232 Th | 40 K | 7 Be | 137 Cs |
|---|---|---|---|---|---|---|
| soil TV NS 1 | <26 | 23.4 ± 1.2 | 31 ± 10 | 382 ± 13 | <23 | <0.27 |
| soil TV NS 2 | 25 ± 9 | 27.9 ± 1.2 | 37.0 ± 1.5 | 479 ± 21 | <40 | 0.6 ± 0.4 |
| soil refinery - tank NS | 11 ± 6 | 10.6 ± 0.4 | 11.0 ± 0.5 | 219 ± 11 | <23 | 0.8 ± 0.2 |
| soil refinery - entrance NS | 20 ± 6 | 18.7 ± 0.5 | 20.3 ± 1.1 | 281 ± 17 | <24 | 1.0 ± 0.4 |
| soil refinery - Oil Plant NS | 20 ± 8 | 16.1 ± 0.7 | 17.5 ± 1.1 | 312 ± 21 | <24 | 0.5 ± 0.3 |
| soil refinery - workshop | 20 ± 11 | 12.5 ± 0.6 | 12.3 ± 1.1 | 273 ± 14 | <20 | 3.2 ± 0.6 |

## 5.2.3. Results

**Table 28. Toxicity and mutagenicity effects of treated soil in bomb craters, Novi Sad**

| Samples | Length of roots in mm | No. of ana-thelophases | No. of aberrat. | Frequency CA % | IC v % |
|---|---|---|---|---|---|
| control $H_2O$ | 35.6 ± 3.4 | *Vicia faba* L. 500 | 1 | 0.2 ± 0.2 | 0 |
| Not diluted | 28.3 ± 1.8 | 500 | 4 | 0.8 ± 0.4 | 20.5 |
| ½ dilution | 29.9 ± 2.4 | 500 | 3 | 0.6 ± 0.2 | 16 |
| ¼ dilution | 34.7 ± 4.1 | 500 | 1 | 0.2 ± 0.2 | 2.5 |
| 1/8 dilution | 40.1 ± 1.5 | 500 | 1 | 0.2 ± 0.2 | 112.6 |
| *Raphanus sativus* subsp. sativus | | | | | |
| control $H_2O$ | 50.6 ± 4.1 | 500 | 1 | 0.2 ± 0.2 | 0 |
| Not diluted | 41.1 ± 2.5 | 500 | 3 | 0.6 ± 0.2 | 18.8 |
| ½ dilution | 49.2 ± 3.1 | 500 | 2 | 0.4 ± 0.2 | 2.8 |
| ¼ dilution | 56.2 ± 2.1 | 500 | 2 | 0.4 ± 0.2 | 11 |
| 1/8 dilution | 62.3 ± 1.3 | 500 | 1 | 0.2 ± 0.2 | 123.1 |
| Zea mays | | | | | |
| control $H_2O$ | 52.4 ± 1.3 | 500 | 1 | 0.2 ± 0.2 | 0 |
| Not diluted | 40.6 ± 3.2 | 500 | 3 | 0.6 ± 0.4 | 22.5 |
| ½ dilution | 46.7 ± 1.0 | 500 | 2 | 0.4 ± 0.2 | 10.9 |
| ¼ dilution | 54.5 ± 1.5 | 500 | 1 | 0.2 ± 0.2 | 104 |
| 1/8 dilution | 62.7 ± 3.1 | 500 | 1 | 0.2 ± 0.2 | 119.6 |

The soil samples from the affected sites in Novi Sad contained 11 ± 6 to 25 ± 9 Bq/kg of $^{238}U$, and most were $^{40}K$ in the range of 219 ± 11 to 479 ± 21 Bq/kg.

In the case of the unmanaged specimens from the Bačka Palanka craters we recorded an IC (inhibitory concentration of the growth of the rootstocks) in the range of 25.6-26.6% and aberrations in the range 0.4-0.6%. In the same site at the edge of the crater, the IC was in the range 21.5-30.3% and aberrations in the range of 0.4-0.8%, and in Novi Sad the IC in the range 18.8-22.5% and aberrations in the range 0.6-0.8%. In the sample diluted to 1/8 we observed growth even stimulated in all three tested species ranging from 12.6 to 23.1% (see Table 28).

### 5.2.4. Conclusion

Regardless of the high concern of above mentioned representatives of relevant countries and their army officials too, only a few studies were done using plants as the bioindicators. It is a pity considering their role in an early warning scheme for a preliminary risk assessment (Lourenço et al. 2016). We may quote the study on lichens (Loppi et al. 2003): "The contribution of the conflict of 1999 to the environmental levels of the uranium in the Balkan area was evaluated by the means of lichens used as biomonitors. The average U concentration found in lichens in the present study was in line with the values reported for lichens from other countries and well below the levels found in lichens collected in areas with natural or anthropogenic sources of U. Measurement of isotopic ratios $^{235}U/^{238}U$ allowed to exclude the presence of depleted uranium. According to these results, we could not detect widespread contamination by depleted uranium in the Balkan area."

Much more studies were done with human cells as an endpoint of testing. Contrary to the above mentioned study with lichens Serbian authors (Milacić et al. 2004) find an increased incidence of the chromosomal aberrations in the blood samples from the residents when analyzed from the three locations in Southern Serbia, that were the target areas for the uranium ammunition in 1999: Bujanovac – Vranje, Kosovo – Strpce and Pljackovica – Przar. A statistically significant increased incidence of the rogue cells and the chromosomal aberrations was found in the blood of the residents of Vranje and Bujanovac. However, when we look closely on these results, we may find that actually the highest chromosomal aberration frequency was found not among the residents, but among the occupationally exposed workers that cannot be connected with U-bombs used in among the bombing of the areas studied (Milacić et al. 2004).

Much larger study was covering the general mortality and/or cancer mortality of Italian veterans deployed in Bosnia-Herzegovina and Kosovo (Capocaccia et al. 2016) reporting: "Mortality rates for the Balkan cohort (71,144 persons) were compared with those of the Italian general population as well as of those of a comparable and unselected control cohort of not deployed military personnel (114,269 persons)… The Balkan cohort experienced mortality rates lower than both the general population (SMR = 0.56; 95% CI 0.51-0.62) and the control group (SMR = 0.88; 95% CI 0.79-0.97). Cancer mortality in the deployed cohort group was half of that from the general population mortality rates (SMR = 0.50; 95% CI 0.40-0.62) and slightly lower if compared with the control group cancer mortality rates (SMR = 0.95; 95% CI 0.77-1.18)."

Our conclusion is that 11 years after the NATO bombing, we had noticed toxicity but not the mutagenity of the soil samples tested.

## 5.3. The Radiation and the Radio-Resistance – Experiments with *Vicia faba* L.

Our previous studies showed that living organisms have a surprisingly high resistance to the radiation. Consequently, the aim of this study was to find limits for surviving an exposure to high doses of the radioactivity for seeds of *Vicia faba* L. using seed stocks of different age and vitality, preserved for decades in our laboratory.

### 5.3.1. Introduction

Resistance to the radiation was already reported in historical cases of human survivors in Goiania (IAEA 1988) and Chernobyl (UNSCEAR 2008), with extreme examples of plants like *Gingko biloga* in Hiroshima (UNITAR 2013). We may confirm it in the case of NPP Jaslovské Bohunice (Murín and Mičieta 2009b). Plants, in particularly, are treated by radiation also for experimental mutagenesis as stressed by "International Symposium on Induced Mutation in Plants" organized by IAEA (Murín and Mičieta 2008). The history of this type of research in former Czechoslovakia is dated back to 1950s and 1960s. The first experiments with barley treated by the radiation for breeding were from 1955 followed by experiments with wheat, beans, rye, barley, oat, soya, rape, hop, cherries, apples, roses, strawberries and grass (c.f. Peška 1980). Breeders in former Czechoslovakia reported an increase of the cereal harvests up to 14–47% with varieties that resulted from the radiation of seeds. Also a better resistance of new varieties to diseases was obtained. For breeding of new mutants of wheat, characterized by a shorter stalk of 15-20 cm, cobalt $^{60}Co$ was used as in the most of the reports (Peška 1980). Therefore, we used the same source of radiation in our experiments.

Šeliga (2006) noted that: "Irradiation of fruit trees resulted in a shorter maturation time, changes in shape and color of the fruits. There are great opportunities of the stimulating radiation in the cultivation of flowers, where various new colors and unusual shapes of flowers are reachable. Irradiation offered new mutants of cereals with a stronger stalk, resistance to lodging and especially better characteristics of flour. Yields per hectare were also increased. It is similar to experiments with rice, which mutants mature earlier and are more resistant. Remarkable results have been obtained, for example after irradiation of potatoes, where the new varieties were obtained." Newer experiments in Slovakia were following authors like Gomez-Pando (2014) by a series of experiments by Gajdošová et al. (2004, 2007), Múdry and Gajdošová (2010) and Hricová et al. (2010, 2011) with *Amaranthus* species.

During 1998 – 2010, twelve generations of mutant lines were established by Hricová et al. (2011). The negative plants (plants with weak seedling growth, non-uniform flowering and seed maturation, with abundant leafiness in the inflorescence area, low size of seeds etc.) were removed and only the plants with positive traits were collected. The WTS (weight of thousand seeds) was recorded and statistically evaluated. Finally, three mutant lines of *A. cruentus* and one line of hybrid K-433 with significantly increased WTS were selected with an obvious tendency to the stabilization of this trait when compared to untreated controls and to the samples of the previous generations.

Finally, Hricová et al. (2011) continued the experiments with induced mutagenesis by experimental irradiation with promising results in the nutrition improvement of amaranths.

However, our experience from field studies showed a different picture in terms of plant response to the irradiation. It is well known from the Chernobyl that plants are not only able to resist high doses radioactivity, but after years may develop a significant adaptability to the radiation (Klubicová et al. 2010, 2011, 2012). We reported the same findings for wild plant species after more than 30 years of radiation contamination caused by a nuclear accident in the NPP Jaslovské Bohunice (Slovakia) for wild plant species (Murín and Mičieta 2009b, Mičieta and Murín 2014). The acute radiation in Novi Sad caused by bombing with low-level uranium dispersed by American weapons was tested by us, with negative results in terms of any genotoxicity again (Mičieta et al. 2010c).

In all previously mentioned reports from the field experiments only actual doses were used and there was not an attempt to find a limiting dose for survival of plants under heavy radiation. Therefore, our aim was to test the vitality of differently aged seeds of *V. faba* L., knowing their sensitivity to the radiation and expecting a higher sensitivity of older ones. Two cultivars were used with different doses of radioactivity in two different experimental settings.

## 5.3.2. Materials and Methods

### *5.3.2.1. Seeds and Their Storage*

In the first series of experiments we were using two cultivars (Inovec and Liber) of *Vicia faba* L. stored for five different periods of time at 4°C in a refrigerator:

1. Inovec = 21 years
2. Inovec = 18 years
3. Inovec = 13 years
4. Liber = 4 years
5. Liber = 2 years

Control: Inovec = 1 year.

In all three samples of cv. Inovec we expected the higher sensitivity confirmed by our very first experiments with old seeds of *Vicia faba* L. (Murín 1988a, b).

In second series of experiments, treated by the higher doses, we used the same two cultivars (Inovec and Liber) of *Vicia faba* L. stored for the three different periods of time at 4°C in a refrigerator:

6. Inovec = 22 years
7. Inovec = 14 years
8. Liber = 2 years

Control 1: Inovec = 1 year

Control 2: Liber = 1 year.

#### *5.3.2.2. Irradiation*

In our experiments was used $^{60}Co$, since it was a main source of the irradiation for the experiments in Czechoslovakia before 1989, as described in experiments by Foltýn, Votava, Velikovský, Ženiščeva and other authors (c.f. Peška et al. 1980). In our first experiment there were three doses: A = 0.5 Gy, B = 1 Gy and C = 2.5 Gy. We may refer to the previous report of authors irradiating *Vicia faba* L. seed with the maximum dose of 2.322 C.kg-1 (Peška et al. 1980), equivalent to 90 Gy. In the second experiment there were three doses again: D = 2.5 kGy, E = 5 kGy and F = 10 kGy.

#### *5.3.2.3. Soaking and Germination*

The seeds were soaked for 2 h in plastic jars of distilled water with a continual air-bubbling. Seeds then germinated in the wet sawdust (Murín and Mičieta 2017a).

#### *5.3.2.4. Tests of Vitality*

Vitality by the rate of a successful germination of seeds and the following length of the roots (from sets of 35 seeds from each sample) were measured after 48 h, 72 h and 96 h. These different recovery times are necessary in this set of experiments to also show an ability of tested cells to recover from the induced treatment in the following cell circles. This effect was closely studied for *Vicia faba* L. in our initial experiments summarized in two monographs (Murín and Mičieta 2001 + 2017b). During the mitotic cycles (checked in recovery times) the repair of DNA damage is taking place with a significant reduction of the evaluated chromosomal aberrations.

#### *5.3.2.5. Cytological Evaluations*

For the cytological evaluation we chose ana-telophases (Bezrukov and Lazarenko 2002). These mitotic figures are simple to evaluate and thus allowed us to make experiments with large numbers of samples under different doses of the radiation and recovery times. Roots of seed samples were fixed in ethanol:acetic acid (3:1), squashed and stained by aceto-orcein. Our aim was to evaluate on average 500 ana-telophases per recovery time for the occurrence of fragments (F), bridges (B) or both (F+B).

#### *5.3.2.6. Statistical Methods*

The statistical analysis of experimental data in all cases was done with analyses of variance (ANOVA) according to Amphelet and Delow (1984) which is based on the Poisson distribution.This guarantees that the confidence level is at least $\alpha=0.05$ also for the probabilities close to zero.

### 5.3.3. Results and Discussion

As expected, the oldest seeds set No. I showed the highest frequency of aberrations, surprisingly under the lowest dose of radiation. The same counts for the set No. II treated by dose A (Table 29). However, the first result (sample IA) in the same Table 29 may be affected by a low number of evaluated ana-telophases. On the other hand, it is a sign of certain fatal impact of radioactivity on the cells when there were minimum chromosomes to be evaluated. The spectrum of obtained results is so wide that they are significant only in comparison between different sets of seeds and between different doses for the same set of equally old seeds like in the sample No. I. An exception for this is the set No. II after 96 hours when comparing the vitality (evaluated by the length of roots) of the plants irradiated by dose A and other two doses, with a significant difference in lower vitality and the highest aberration rate although it was treated by the lowest radiation dose. Other tendencies in the set No. II were insignificant (Table 29). Also other doses (III, IV and V) are in vitality and chromosomal aberration practically insignificant. Generally, the impact of different radiation doses was insignificant between different doses, although some tendencies were visible.

**Table 29. Vitality and frequency of chromosomal aberrations for *Vicia faba* L. seeds sets different in cultivars, time of storage and doses of irradiation**

| Cultivar and time of storage | Treatment dose (Gy) | Length of roots (mm) | | | Chromosomal aberrations after 96 hours | | | | |
|---|---|---|---|---|---|---|---|---|---|
| | | after 48 hours | after 72 hours | after 96 hours | Number of evaluated ana-telophase | F | B | F+B | Frequency of aberrations (%) |
| Inovec (21 years) | 0.5 | 5.78 ± 0.92 | 14.67 ± 3.60 | 18.44 ± 4.54 | 84 | 3 | 3 | 0 | 7.14 ± 1.44 |
| | 1.0 | 4.00 ± 0.58 | 9.67 ± 2.87 | 13.22 ± 4.38 | 338 | 3 | 9 | 0 | 3.55 ± 0.76 |
| | 2.5 | 6.67 ± 1.14 | 14.00 ± 3.82 | 20.89 ± 5.98 | 364 | 7 | 5 | 1 | 3.57 ± 1.33 |
| Inovec (18 years | 0.5 | 4.22 ± 0.74 | 7.89 ± 2.49 | 9.89 ± 3.81 | 181 | 7 | 5 | 0 | 6.63 ± 2.01 |
| | 1.0 | 5.67 ± 1.04 | 13.00 ± 3.89 | 19.22 ± 6.34 | 286 | 1 | 1 | 0 | 0.70 ± 0.22 |
| | 2.5 | 4.56 ± 0.84 | 6.55 ± 2.43 | 8.44 ± 4.20 | 122 | 1 | 1 | 0 | 1.64 ± 0.45 |
| Inovec (13 years) | 0.5 | 9.89 ± 0.96 | 22.44 ± 2.48 | 36.00 ± 4.15 | 500 | 3 | 3 | 0 | 1.20 ± 0.38 |
| | 1.0 | 7.78 ± 0.74 | 18.33 ± 3.14 | 25.89 ± 4.87 | 500 | 3 | 2 | 0 | 1.00 ± 0.07 |
| | 2.5 | 7.56 ± 0.73 | 20.22 ± 3.10 | 33.00 ± 5.57 | 500 | 1 | 0 | 0 | 0.20 ± 0.11 |
| Liber (4 years) | 0.5 | 8.89 ± 0.75 | 24.00 ± 2.92 | 34.89 ± 5.01 | 500 | 1 | 1 | 0 | 0.40 ± 0.08 |
| | 1.0 | 9.56 ± 0.88 | 23.89 ± 2.75 | 35.89 ± 5.04 | 500 | 0 | 1 | 0 | 0.20 ± 0.06 |
| | 2.5 | 8.11 ± 0.70 | 22.67 ± 3.61 | 38.67 ± 6.25 | 500 | 2 | 1 | 0 | 0.60 ± 0.28 |
| Liber (2 years) | 0.5 | 15.33 ± 0.50 | 33.78 ± 2.23 | 52.00 ± 4.01 | 500 | 3 | 3 | 0 | 1.20 ± 0.64 |
| | 1.0 | 13.89 ± 0.75 | 30.33 ± 2.02 | 46.22 ± 3.84 | 500 | 1 | 1 | 0 | 0.40 ± 0.22 |
| | 2.5 | 11.89 ± 0.84 | 26.33 ± 1.28 | 42.00 ± 2.29 | 500 | 0 | 1 | 0 | 0.20 ± 0.06 |
| control | | 14.00 ± 0.80 | 28.89 ± 0.93 | 44.44 ± 2.26 | 500 | 2 | 0 | 0 | 0.40 ± 0.09 |

**Table 30. Vitality and frequency of chromosomal aberrations for *Vicia faba* L. seeds sets different in cultivars, time of storage and doses of irradiation in reduced mode**

| Cultivar and time of storage | Treatment dose (kGy) | Length of roots (mm) | | | Chromosomal aberrations after 96 hours | | | | |
|---|---|---|---|---|---|---|---|---|---|
| | | after 48 hours | after 72 hours | after 96 hours | Number of monitored ana-telophase | F | B | F+B | Frequency of aberrations (%) |
| Inovec (22 years) | 2.5 | 4.67 ± 1.22 | 18.55 ± 3.36 | 24.78 ± 4.73 | 500 | 2 | 1 | 0 | 0.60 + 0.12 |
| | 5.0 | 4.89 ± 2.02 | 12.00 ± 4.40 | 15.67 ± 5.60 | 328 | 6 | 0 | 0 | 1.83 + 0.08 |
| | 10 | 6.67 ± 1.37 | 17.78 ± 3.28 | 23.89 ± 4.65 | 281 | 2 | 1 | 0 | 1.07 + 0.10 |
| Inovec (14 years) | 2.5 | 10.56 ± 1.09 | 27.89 ± 0.87 | 39.11 ± 1.01 | 500 | 1 | 2 | 0 | 0.60 + 0.05 |
| | 5.0 | 11.44 ± 1.19 | 30.44 ± 3.13 | 41.00 ± 4.22 | 448 | 4 | 0 | 0 | 0.89 + 0.26 |
| | 10 | 8.89 ± 0.68 | 21.55 ± 1.89 | 29.44 ± 2.63 | 499 | 4 | 1 | 0 | 1.00 + 0.02 |
| Liber (2 years) | 2.5 | 14.89 ± 0.54 | 28.44 ± 1.78 | 38.56 ± 3.56 | 444 | 3 | 0 | 0 | 0.70 + 0.12 |
| | 5.0 | 12.00 ± 1.53 | 22.67 ± 3.30 | 32.89 ± 4.67 | 451 | 5 | 0 | 0 | 1.11 + 0.43 |
| | 10 | 14.56 ± 0.87 | 27.44 ± 1.34 | 36.89 ± 2.06 | 390 | 4 | 2 | 0 | 1.54 + 0.24 |
| Inovec (1 year) | | 12.54 ± 0.56 | 27.09 ± 1.48 | 36.64 ± 2.86 | 500 | 2 | 0 | 0 | 0.40 + 0.08 |
| Liber (1 year) | | 16.33 ± 1.03 | 29.78 ± 1.72 | 38.44 ± 3.24 | 336 | 2 | 0 | 0 | 0.60 + 0.11 |

These results might be heavily affected by individual variability, as we already proved in our earliest experiments (Murín 1988a, b). On the other hand, the tendency of improvement in frequency of aberrations from worse to the best one from No. I to No. V is so clear that the last one (following radiation treatment in the lowest dose) showed even a higher vitality than the control after 96 hours. This can be seen as a certain stimulation effect. It is the same with the length of roots showing the vitality of seeds under different age and doses of irradiation. Same were observations of Geras'kin et al. after γ-irradiation of barley seeds with doses in the range of 2-50 Gy. Plants' growth stimulation occurs in dose range 16-20 Gy. It was shown that increased size of seedlings after irradiation with stimulating doses was due to the enhancing pace of development rather than an earlier germination. The activity of the majority of the enzymes studied increased in the range of doses that cause stimulation of seedlings development.

When Gajdošová et al. (2004) irradiated *Amaranthus* with 0.5, 0.75, 0.9, 1.2 and 1.5 kGy they found no differences in *Eco* RI and *Hae* III restriction patterns of the control plants and irradiated M3 plants in a restriction analysis of chloroplast DNA based on PCR-RFLO analysis of the flaking regions between the genes *trnS* and *psbC.*

The *Eco* RI restriction profiles of control plants obtained from non-irradiated seeds and the 23 plants of M3 generation grown from irradiated seeds consisted of 1 restriction fragment of an approximate size of 2,000 bp. On the contrary, the *Hae* III patterns consisted of 5 restriction fragments of 1 300, 500, 210, 100, and 70 bp, respectively, uniformly occurring in all the samples compared. However, we expected that *V. faba* seeds can sustain much higher doses relatively unharmed when irradiated in a dry (dormant) stage as was already observed with wild plant species (Mičieta and Murín 2014).

**Table 31. Vitality of the same seeds sets as in Table 30 expressed by their germination**

| Germination (%) | | | | | | | | | | | | | | | |
|---|---|---|---|---|---|---|---|---|---|---|---|---|---|---|---|
| kGy | after 48 hours | | | | | after 72 hours | | | | | after 96 hours | | | | |
| | IN 1 yr. | LI 1 yr. | IN 22 yrs | IN 14 yrs | LI 2 yrs | IN 1 yr. | LI 1 yr. | IN 22 yrs | IN 14 yrs | LI 2 yrs | IN 1 yr. | LI 1 yr. | IN 22 yrs | IN 14 yrs | LI 2 yrs |
| 2.5 | 100 | 100 | 78 | 100 | 100 | 100 | 100 | 89 | 100 | 100 | 100 | 100 | 89 | 100 | 100 |
| 5.0 | | | 44 | 100 | 100 | | | 78 | 100 | 100 | | | 78 | 100 | 100 |
| 10 | | | 89 | 100 | 100 | | | 89 | 100 | 100 | | | 89 | 100 | 100 |

Cultivar Inovec = IN, cultivar Liber = LI.

The second experiment did not show any significant differences between the doses of radiation on seeds of a different age. On the other hand, the tendency of improvement in vitality from the oldest (No. VI) to the youngest seed sets and from the highest to the lowest doses of radiation was more uniform. The wide range of particular results still made the most of comparisons between doses inside of particular seed sets insignificant. We may confirm that with younger seeds sets their vitality improved clearly up to the control level (Table 30).

Germination of seeds was also evaluated in this experiment. Gajdošová et al. (2004) reported that a dose of 500 Gy resulted in 67% growth reduction of *A. cruentus*, resp. 55% of genotype K-433. As we may see from the Table 31 for seeds of *Vicia faba* L. only in the case of 22 years old cv. Inovec all doses had a significant impact. Others (17 years old cv. Inovec and 2 years cv. Liber) had 100% germination rate in all recovery times. Unexpectedly, the lowest (D) and the highest doses (F) had an impact on the germination of the oldest seeds but not as strong as the dose E for the set No. VI (Table 31).

It is important to know that the same mechanisms of adaptation and radio-resistance were observed in agricultural plants populations too (Klubicová et al. 2010 + 2011).

## 5.3.4. Conclusion and Future Perspectives

Based on the previous reports of other authors and our experiments (Murín and Mičieta 2017a), we may conclude:

1. A significant tolerance of chosen plant species to high doses of radiation was found. That is promising in case of any accidental radiation contamination that may occur by human activities.
2. Even with the highest doses of 10 kGy of $^{60}Co$ we did not find a limit of the survival even for very old seeds sets (22 years).
3. Furthermore, our experiments showed that irradiation of cultural plant species has to be well balanced as their response (as in this case of *Vicia faba* L.) is heavily affected by individual variability.

We must stress that our results showed that plants could resist very high doses of irradiation by $^{60}Co$ up to 10 kGy, especially in the dry stage of their seeds.

*Chapter 6*

# GENOTOXICITY OF OIL POLLUTION ON SOME SPECIES OF KUWAITI FLORA

## 6.1. INTRODUCTION

Environmental pollution resulting from burning of the oil wells in the Kuwaiti desert during the Gulf war, has been under our genotoxic investigation using plant species grown there as bioindicators. The cytogenetic assays performed to detect the mutagenicity of the oil pollution included pollen abortion analysis, pollen mother cell (PMC) analysis and mitotic analysis for chromosomal aberrations.

Sterility of the anthers, the presence of meiotic irregularities and inhibition of the growth were the main indicators of disturbance. An unusual cytokinesis, which led to the formation of four pollen grains without the usual cell plate formation, was also observed.

Out of 58 native plant species, 15 were selected as bioindicators according to certain criteria. Some were sensitive and could be used for the short-term pollution testing and others were resistant and could only be used for the long-term pollution testing.

During the Gulf war in 1990, more than 700 oil wells were set alight, resulting in the burning of 4-6 million barrels of crude oil daily. These fires were daily emitting thousands of tons of nitrogen and sulfur oxides and thousands of tons of soot. Up to 20 million tons of unburnt crude oil leaking around the oil wells were forming lakes. The acid gases and the oil lakes had a tremendous adverse effect on the environment including the flora and fauna of Kuwait. Up to 8 million barrels of crude oil were released into Persian Gulf.

Bioindicators change their occurrence, vitality and other responses under the influence of the environmental conditions (Kovacs 1992). Each bioindicator responds to environmental change as it would respond to a specific stimulus. The absorbed stimuli bring reactions that provide information on both the changes and the level of pollution. The data provided by bioindicators can be used to estimate the degree of environmental impact and its potential danger for other living organisms.

The present investigation was based on the wild flowering plants of Kuwait as bioindicators. Flowering plants were collected after the war in 1992 and 1993 for the screen of the effects of the oil pollution on their genetic material. The main goals of this work were:

1. Determination of the mutagenicity of oil pollution at some species of flowering plants.
2. Selection of species from the Kuwaiti flora to be used also in the future as indicators for detecting environmental pollution.

## 6.2. Materials and Methods

### 6.2.1. Plant Sources

Plants were collected from several locations in Kuwait. Taxonomic identification of the herbarized material was kindly done by Professor L. Boulos from University of Kuwait. Young flowers and flower buds at different developmental stages were collected from at least 10 specimens of every species. The samples were fixed in Carnoy's fixative, and stored in a refrigerator at 4°C until used for study. Seeds were collected and kept in polyethylene bags.

The bioindicator plants for this study were selected according to the following criteria:

1. Distribution of plant species.
2. Ploidy state of chromosomes.
3. Positive end point with bioassays.
4. Viability of pollen grains under the normal climatic conditions.
5. Blooming seasons during the year.

The degree of fitness of every species to be used as a bioindicator according to the bioassays performed was scaled as: +++ = very good, ++ = good, + = fair, ? = difficult to obtain and -- = not tested.

### 6.2.2. Cytogenetic Assays

#### *6.2.2.1. Pollen Mother Cell (PMC) Analysis*

Young buds were taken out from the fixative and macerated in a mixture of ethanol and hydrochloric acid (1:1) for 5-10 min. and then washed with distilled water. The anthers were squashed and stained in an acetic-orcein reagent, on a glass slide. About 300 cells, collected from 10-20 flower buds, were analyzed. Meiotic configurations were observed and recorded.

#### *6.2.2.2. Pollen Abortion (Sterility) Analysis*

Mature buds were used. The preparation of slides was carried out in the same manner as for PMCs. In some cases, aniline blue was used for staining. The ability for staining, the regularity of shape and size were the criteria for pollen abortion analysis. At least 1000 pollen grains were observed in at least 2-3 inflorescences for every plant species.

#### *6.2.2.3. Mitotic Analysis*

Seeds were germinated in pots with the wet soil or at the Petri dishes covered with the wet filter paper. Young roots were excised and fixed in

3:1 mixture of an absolute ethanol and acetic acid, respectively, and kept at 4°C until used. For slide preparation, roots were macerated and stained. Mitotic analysis was done from about 300 cells and mitotic configurations were observed and recorded.

#### *6.2.2.4. Tetrad Analysis*

Scoring of micronuclei in tetrads was carried out according to the staining ability and the appearance of micronuclei in a pollen grain cell. The criteria of phytotoxicity include visible morphological changes and presence/absence of the plant cell division.

#### *6.2.2.5. Ecotoxicological Analysis*

The germinating capacity was tested in some species by germinating the seeds for different time durations. Clastogenicity was determined by analyzing the root tips of the germinated species for the structural chromosomal changes.

#### *6.2.2.6. Vicia sativa Test (cf. Murín 1984)*

A water extract of the crude oil from Burgan lake was made by mixing 3 parts of water to 1 part of oil. The extraction lasted for either 24 hrs or 3 weeks with some shaking every day. The seeds of *Vicia sativa* L., var. Fatima were soaked in the extract for 24 h and then germinated in Petri dishes on filter papers wetted with the same extract. After 72 h of germination the roots of the seedlings were measured and root tips were used for scoring chromosome aberrations.

To increase the content of toxic substances in the extract, small amount of organic solvents were used. The seeds were then soaked in the following emulsion: 5 ml xylene + 5 ml of Tween SO + 5 ml crude oil in 985 ml of distilled water. Extraction lasted for 24 h and then soaking of the seeds started. After 24 h the seeds were germinated in the Petri dishes on the filter papers wetted with the same emulsion for another 72 h. After measurement of the roots, the root tips were used for the scoring of chromosome aberrations.

### 6.2.3. Locations of Sampling

Our main object was Burgan area – the oil field where the heavy fire raged. Qurtoba, Khiran and Andalus, 25-30 km away from Burgan oil fields, were used as control locations.

### 6.2.4. Statistical Analysis

Between the samples from the control area and polluted areas were statistical analysis conducted following the procedure of Amphlett and Delow (1984), which is based on the Poisson distribution.

## 6.3. Results

### 6.3.1. Pollen Mother Cell Analysis (PMC)

The percentage of abnormal meiotic cells in *Plantago boissieri* was higher in the location Burgan than the location Qurtoba during 1992 (Table 33), whereas the percentage was low at all locations in the following year.

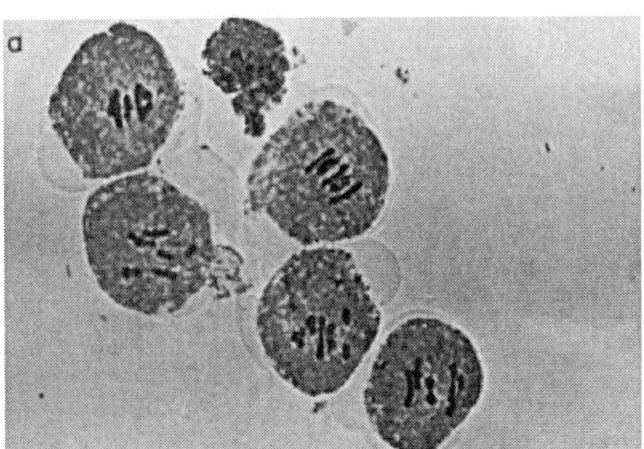


Figure 29. Spore mother cell in *Picris babylonica* with 4 bivalents plus 1 pair of lagging chromosomes.

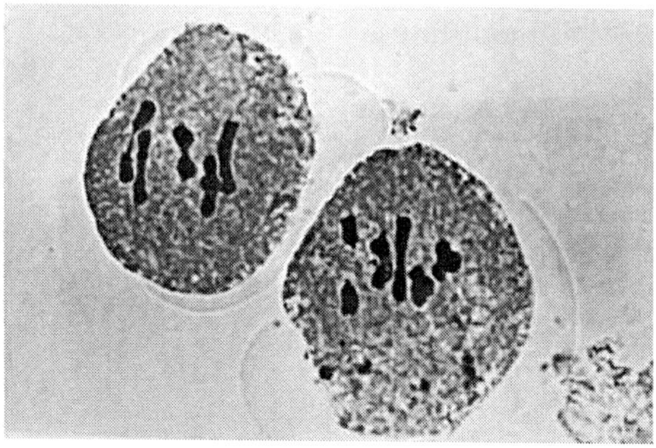

Figure 30. Spore mother cells in *Picris babylonica* with 4 bivalents plus 1 pair of lagging chromosomes (in detail).

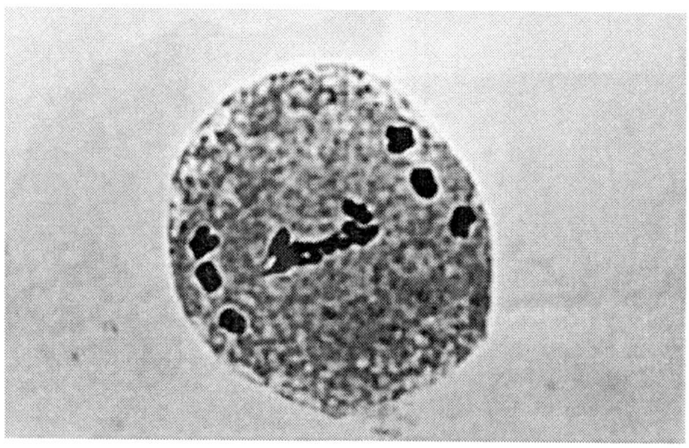

Figure 31. Metaphase I in *P. babylonica* in detail.

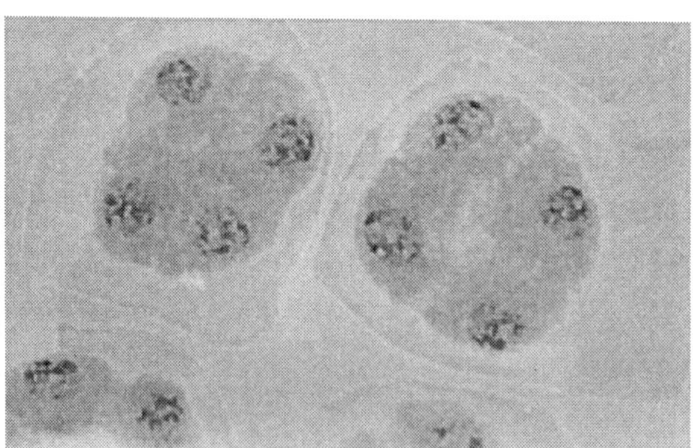

Figure 32. Beginning of cytokinesis in spore mother cells of *Plantago boissieri*.

However, for *Launaea mucronata* and *Launaea capitata* the percentage was within the standard range, except an adverse effect in the meiotic analysis of *Launaea mucronata* which has no significant value (Table 34 and 35).

*Picris babylonica* showed a marked increase in abnormal meiotic cells in buds collected from Burgan in 1993, opposite results of tetrad analysis were under high standard error deviation (Table 36). Metaphase I in *P. babylonica* showed 5 bivalents in a majority of samples, but metaphases with 4 bivalents plus 1 pair of lagging chromosomes were also seen in considerably large numbers (Table 37).

The meiotic analysis showed various kinds of abnormalities such as occurrence of fragments, bridges, lagging chromosomes, micronuclei, etc. (Figures 29-32).

An unusual type of cytokinesis (division of cytoplasm) was also observed (Figure 31-32). The figures show that the formation of four pollen grains occurred by dividing the cytoplasm into four nearly equal portions but without the formation of cell plate. Also, an abnormal cytokinesis was observed where the cytoplasm was divided into several unequal portions (3-8), each with a nucleus. Some chromatin material was found in form of chromosomes distributed unequally in the cytoplasmic portions.

### 6.3.2. Pollen Abortion (Sterility) Analysis

In 1992, some anthers of location Qurtoba showed a normal percentage of abortive pollen grains, 5.3% (Table 33), while other anthers were full of abortive pollen grains (8 buds from 3 different inflorescences showed 100% abortive pollen grains). The same result was found from specimens collected from the Burgan area. Some anthers showed low percentages of abortion, 1.8% (Table 32), while other anthers were full of abortive pollen grains.

In 1993, anthers from the location Qurtoba did not show any normal pollen grains. The examined anthers were full of abortive pollen grains.

Some anthers from the location Burgan showed normal percentage of abortion, 4.0% (Table 33) while other anthers were full of abortive pollen grains (18 buds of 6 different inflorescences showed complete abortive pollen grains).

**Table 32. The percentage of abnormal cells and abortive pollen grains in *Plantago boissieri* obtained from the pollen test, meiotic and mitotic tests**

| | **1992** | | | **1993** | | |
|---|---|---|---|---|---|---|
| | Qurtoba | Andalus | Burgan | Qurtoba | Andalus | Burgan |
| tec | 5230 | - | 1523 | - | 5938 | 1533 |
| PA | | | | | | |
| % | 5.3 | - | 1.8 | - | 5.0 | 4.0 |
| tec | 953 | - | 3076 | 1743 | 2810 | 4928 |
| MEA | | | | | | |
| % | 2.3 | - | 4.3 | 1.4 | 2.5 | 1.7 |
| tec | 391 | - | 337 | 480 | - | 473 |
| MIA | | | | | | |
| % | 1.5 | - | 0.1 | 0.4 | - | 0.2 |

PA = pollen abortion, MEA = meiotic analysis, MIA = mitotic anal.
tec = total examined cells, % = % of abnormality.

**Table 33. The percentage of abnormal cells and abortive pollen grains in *Launea mucronata* obtained from pollen test, meiotic and mitotic tests**

| | **1992** | | **1993** | |
|---|---|---|---|---|
| | Q/A/R | Burgan | Q/A/R | Burgan |
| tec | 3596 | 1003 | 1500 | 1000 |
| PA | | | | |
| % | 2.0 | 0.7 | 0.0 | 0.6 |
| tec | 452 | 393 | 452 | 990 |
| MEA | | | | |
| % | 0.0 | 0.8 | 0.0 | 0.0 |
| tec | 307 | 304 | 352 | 359 |
| MIA | | | | |
| % | 1.3 | 1.6 | 0.8 | 0.5 |

PA = pollen abortion, MEA = meiotic analysis, MIA = mitotic anal. tec = total examined cells, % = % of abnormality,Q/A/R = Qurtoba, Andalus, Remathia.

**Table 34. The percentage of abnormal cells and abortive pollen grains in *Launaea capitata* obtained from pollen test, meiotic and mitotic tests**

| | 1992 | | 1993 | |
|---|---|---|---|---|
| | Qurtoba | Burgan | Qurtoba | Burgan |
| tec | 3010.0 | 3991.0 | 3269.0 | 3368.0 |
| PA | | | | |
| % | 2.1 | 4.4 | 1.4 | 1.6 |
| tec | 41.0 | 258.0 | 278.0 | 488.0 |
| MEA | | | | |
| % | 4.4 | 0.0 | 0.0 | 0.0 |
| tec | 375.0 | 436.0 | 535.0 | 494.0 |
| MIA | | | | |
| % | 0.5 | 0.2 | 0.1 | 0.0 |

PA = pollen abortion, MEA = meiotic analysis, MIA = mitotic anal. tec = total examined cells, % = % of abnormality.

The percentage of abortive pollen grains from *Launaea mucronata* and *Launaea capitata* were within normal range (Table 33 – 34).

### 6.3.3. Mitotic Analysis

**Table 35. The percentage of abnormal cells and abortive pollen grains in *Picris babylonica* obtained from tetrad analysis, meiotic and mitotic tests**

| | 1992 | | | 1993 | | |
|---|---|---|---|---|---|---|
| | Qurtoba | Khiran | Burgan | Qurtoba | Khiran | Burgan |
| tec | 407 | - | 587 | 883 | 284 | 1509 |
| TA | | | | | | |
| % | 5.15 | - | 2.72 | 1.36 | 1.4 | 6.22 |
| tec | 1148 | - | 1730 | 758 | 865 | 1806 |
| MEA | | | | | | |
| % | 3.83 | - | 1.56 | 2.77 | 0.57 | 8.05 |
| tec | 398 | - | 337 | 326 | - | 359 |
| MIA | | | | | | |
| % | 1.00 | | 2.0 | 0.0 | - | 2.5 |

TA = tetrad analysis, MEA = meiotic analysis, MIA = mitotic anal. tec = total examined cells, % = % of abnormality.

The percentage of abnormal cells was low for the species from all the locations studied (Table 35-36).

**Table 36. The analysis of metaphase I in *Picris babylonica***

| | **1992** | | **1993** | |
|---|---|---|---|---|
| | Qurtoba | Burgan | Qurtoba | Burgan |
| tec | 116 | 226 | 250 | 308 |
| BIa | 96 | 164 | 200 | 175 |
| BIb | 5 | 62 | 50 | 48 |
| BIc | 15 | - | - | 85 |
| % | 17.25 | 27.43 | 20.00 | 43.18 |

TEC = total examined cells, % = % of abnormality, BIa = (n=5) bivalents, BIb = 4 bivalent + 1 separated bivalent. BIc = 5 bivalent + fragment.

## 6.3.4. List of Suitable Bioindicators

**Table 37. List of species of Kuwaiti native flora selected as suitable bioindicators**

| **Species** | **Pollen abortion test** | **Meiotic test** | **Mitotic test** |
|---|---|---|---|
| *Allium sindjarensis* Boiss. & Hausskn. | ? | ? | +++ |
| *Arnebia decumbers* (Vent.) Coss. & Kral. | ++ | ? | ? |
| *Bassia muricata* (L.) Aschers. | ++ | -- | -- |
| *Bassia eriophora* (Schrad.) Aschers. | ? | ? | +++ |
| *Brassica tournifortii* Gouan. | ++ | ++ | ? |
| *Erodium lacinatum* (Cav.) Willd. | +++ | ? | ? |
| *Helianthemum lippii* (L.) Dum. Cours. | ++ | ? | ? |
| *Launaea capitata* (Spreng.) Dandy | +++ | +++ | +++ |
| *Launaea mucronata* (Forssk.) Muechler. | +++ | +++ | +++ |
| *Neurada procumbens* L. | +++ | ? | ? |
| *Picris babylonica* | ++ | +++ | ++ |
| *Plantago amplexicaulis* Cav. | ++ | ++ | ++ |
| *Plantago boissieri* & Born M. | +++ | +++ | +++ |
| *Plantago lanceolata* L. | ++ | ++ | ++ |
| *Reseda arabica* Boiss. | ++ | ? | ? |

The degree of fitness of every species to be used as a bioindicator according to the bioassays performed was scaled as: +++ = very good, ++ = good, + = fair, ? = difficult to obtain.

Table 37 represents the list of plant species that were collected from all locations. Fifteen out of fifty eight species were selected as bioindicators for the study of environmental pollution. The scale shown in Table 37 indicates the fitness of every species according to the bioassay.

### 6.3.5. Seed Germination

**Table 38. The capacity of seed germination for some collected plant species**

| Species | Burgan 1992 | Qurtoba 1993 |
|---|---|---|
| *Plantago boissieri* | | |
| 96 h | 30% | 20% |
| 120 h | 31% | 21% |
| 144 h | 32% | 24% |
| Clastogenicity | 2% | 1% |
| *Plantago amplexicaulis* | | |
| 96 h | 24% | 88% |
| 120 h | 24% | 88% |
| 144 h | 24% | 88% |
| Clastogenicity | n.e. | n.e. |
| *Launaea capitata* | | |
| 96 h | 71% | 83% |
| 120 h | 71% | 83% |
| 144 h | 71% | 83% |
| Clastogenicity | n.e. | n.e. |
| *Medicago laciniata* | | |
| 96 h | 10% | 7% |
| 120 h | 13% | 9% |
| 144 h | 16% | 12% |
| Clastogenicity | 0.3% | 0.3% |

n.e. = not examined.

The germination capacity was tested for the seeds of several species as shown in Table 38. The data showed a variable reduction effect.

### 6.3.6. *Vicia sativa* Analysis

This test has shown that the extraction of crude oil by water must be longer than 24 h to get enough of water soluble toxicants for detection of phytotoxic effects. This process may simulate the situation in the field in the case of pollution by crude oil or its residues. The data in Table 39 show a clear inhibition of root growth for the treated seeds and seedlings. Some increase in toxic substances during the 24 h extraction was found by using some of the organic solvents, though the solvents themselves cause an inhibitory effect. Prolongation of extraction can increase the inhibitory effect. As to the clastogenic effect the mitotic and chromosomal observations have not shown any increase when compared with the controls. This suggests that alternative methods and times of extraction should be adopted for genotoxic testing.

**Table 39. Results of *Vicia sativa* analysis after treatment with water soluble crude oil extracts and extract made with organic solvents in water**

| Treatment | Root length (mm) | Clastogenicity |
|---|---|---|
| 24h extraction | | |
| water/oil 3:1 | 54.2 + 2.3 | 0.3% + 0.3% |
| control | 53.4 + 3.8 | 0.3% + 0.3% |
| 3 weeks extraction | | |
| water/oil 3:1 | 49.6 + 2.7 | 0.0% + 0.0% |
| control | 67.8 + 3.5 | 0.3% + 0.3% |
| 24h extraction | | |
| mixture | 20.9 + 2.9 | 0.3% + 0.3% |
| mix. control | 29.5 + 2.9 | 0.3% + 0.3% |

Mixture: 5ml of xylene, 5 ml of Tween 80, 5ml of crude oil, added and emulsion made in 985 ml of dist. water, water mix. control: solvents + 985 ml. dist. water.

## 6.4. DISCUSSION

The distribution of species in urban and industrial habitats is essential for studying many alterations in environmental quality (Murín 1995). In 1992, *P. boissieri* was found only in two of the areas selected for this study in Kuwait, whereas in 1993, the same species was found in three locations. This might be an indication of an acute toxicity response that occurred shortly after the oil pollution during the Gulf war in 1990-91. In long-term studies of a site, changes in species numbers may provide an acceptable basis for characterizing the response to pollution (Kovacs 1992). Unfortunately, the conditions in Kuwait after the liberation were not safe due to infestation of the desert with mines. Therefore, it was not possible to screen and monitor the distribution of species under investigation in more areas. Burton (1989) found that varied sensitivity allows some species to compete more successfully in stressed conditions, whereas others are eliminated because of acute toxicity responses or impaired growth.

Pollen grains show several characteristics which allow them to serve as good indicators of environmental mutagenesis (Wolters and Martens 1987). Pollen grains are not only sensitive to genetic changes, but also to non-mutagenic environmental stresses. For instances, moderate temperature or drought stress can totally disrupt the pollen development (Mulcahy 1981). The effect of air pollution, resulting from the burning of oil wells and the resulting extremely high temperature could be observed in the inviability of pollen grains. The presence of anthers full of abortive pollen grains reflect the sterility of these plants. This condition was observed at many locations indicating the severity of polluted areas. The 100% abortive pollen grains of many anthers from different plants provide the evidence that the plant species were under severe environmental stress. Long-term investigations would reveal the extent of genetic damage to the pollen system and whether it is inherited. The viability of pollens is dependent upon a large number of loci. Pfahler (1975) has summarized a series of studies which indicate that not only even very small deletions in virtually any part of the genome will induce pollen abortion, but also that any point mutations will have equally lethal effects upon the male gametophyte. On the cellular and subcellular level, the effects of petroleum and petroleum products on animal or plant cells have not been fully investigated. The water-soluble fractions (WSFs) of fuel oil have been shown to be particularly toxic to the embryonic stages of marine zooplankton (Byrne 1989). Renzoni (1975) observed abnormalities in the development of the larvae of the mollusc *Mulinia lateralis* exposed to WSFs of Kuwait, Nigerian and Prudhoe Bay crude oils. Deleterious effects on the fertilization of some marine eggs were also observed (Nicol et al. 1977, Falk-Petersen 1979, Koster and Van den Biggelaar 1980). The effects noticed on the development of these marine organisms confirm the sensitivity of the cell furrow exposure to soluble petroleum hydrocarbons. The plant species under investigation showed varied percentage of meiotic abnormalities. However, none of them seems to be significant. Nonetheless, several observations may reflect the effect of oil pollution on the plant cells, for example, an increase in some abnormal meiotic irregularities. This condition may well be attributed to stress (Shull and Meazel 1977). Increased presence of micronuclei is also an indication of genotoxicity for many species (Sandhu et al. 1991). In our studies, *P. babylonica* showed considerable increase in micronuclei which indicates the genotoxic effect of environmental pollution on its genome. The behavior of the cells at cytokinesis which lead to the formation of more than four pollen grains each with a nucleus, also indicates environmental pollution stress in this species.

Cytokinesis in *Launaea* species and *Picris babylonica* was demonstrated in a way other than that of the formation of cell plate. It is not clear so far whether this condition is related to the species specificity or to the effects of stresses on cell plate formation as many chemicals are known to be inhibitors of cell plate formation (Bonsignore and Helper 1985, Helper and Bonsignore 1990).

Mitotic analysis was used as an endpoint for genotoxicity of oil pollution in plant species under investigation as in *Vicia sativa* root tips treated with oil-water extract. Clastogenicity was not clearly demonstrated by this assay.

The growth of the root tips of *V. sativa* treated with oil-water extract was recorded. The inhibition of root growth was noticed in seedlings treated with the extract when the extraction lasted for three weeks. Crude oil application was found to delay the rate of germination of *Zea mays* seeds after the seventh day of application (Amadi et al. 1993).

Fifteen out of fifty-eight native species were selected as bioindicators (Table 38). The selection was made according to the criteria mentioned in the Materials and Methods section. The selected species could react positively with performed bioassays. The sensitivity as bioindicator was variable among species. *P. boissieri* seemed to be sensitive and could be used for short-term exposure to pollutants, while *Launaea* species seemed to be more resistant and could be used for long-term exposure to pollutants. More investigations on cultivated species are needed to avoid the limitation of rain dependent desert flora. Wild flowering plants seem to be good systems for detecting genetic damage due to environmental stresses. It is an effective and low-cost method that could be used in many pollution events.

The bioassays used are found to be demonstrative and can reflect the genetic variation in plant species.

Screening for special characteristics in any species will enhance the value of pollution assessment. *Picris babylonica* has clear tetrads that provide an easy way to detect micronuclei. The presence of micronuclei is an indicator of mutagenicity for many pollutants (De Simone et al. 1992).

**Table 41. Benefits and limitations of phytoremediation**

| Benefits | Limitations |
|---|---|
| Realizable "*in situ*" | Problematic realization at shallow frame substrates (soils), rapid stream and underground waters |
| Passive, undemanding | Possible high ecotoxicity for plants |
| Solar energy utilization | The limitation of the transfer of large volumes associated with other methods of remediation |
| Just 10% up to 20% costs in comparison to manual or chemical elimination | Significantly slower effect in comparison to physical elimination or utilization of chemo–physical methods |
| Minimal environmental damage, simulates natural conditions | Effective only for an average hydrophobic contaminants |
| Well acceptable by public | The high degree of uncertainty, or the lack of knowledge of the toxicity and bioactivity of degradation products |
| None, or a little air and hydro emissions | Contaminants can be mobilized in groundwater |
| No secondary waste production | The potential contamination of animals food chain |
| Convenient for various types of pollutants, remnants that are usable directly on site in subsequent processes | Many unknown regulators |
| Leading up to the mineralization of organic compounds | Possible influence to the course of the decontamination by the change of environmental conditions, e.g., with presence and the concentration of other xenobiotics |

The main tool for phytoremediation is the use of vegetation "*in situ*" with the targeted treatment of the contaminated components of the environment (Schnoor 2000). It is accepted as potentially promising clean-up technology (Marques et al. 2009). For this purpose also ornamental plants can be used (Sulleiman and Bhat 2003).

Plants demonstrate the ability to decontaminate polluted sites (even in such a heavy stage like with the soils polluted by oil) effectively and economically (Amadi et al. 1993). Plants can adsorb various inorganic and organic contaminants from water and soil. Adsorption, metabolic transformation, deposition and accumulation are suitably and potentially useful for the remediation of contaminated habitats. Based on the current knowledge, it applies the best in the case of surface contamination and is particularly effective for hydrophobic substances – nitro compounds and chlorinated solvents (Glass 1999).

All phytoremediation systems: hydraulic barriers, vegetation cover, rhizofiltration, phytoextraction, phytostimulation, phytovolatilization, phytotransformation (Schnoor 1997, Mičieta and Polák 2000) are based on mainly at the strategy of phytostabilization and phytodecontamination (Table 40). However, their most effective utilization needs targeted monitoring of the particular contaminated site and consequently, with and knowledge from the basic research, a suitable range of species can be selected (Jordahl et al. 1997). Still there are, apart from benefits, some notable limitations (Table 41).

In phytoremediation, an ability of plants to intake and metabolite the contaminants that are biologically reachable is crucial. The most important mechanisms are: root absorption, transport in plant tissues, accumulation in the form of non-phytotoxic metabolites, enzyme producing supporting microbial and biochemical transformation, increased mineralization in the rhizosphere and absorption by the leaf surface from the atmosphere. The procedure started by phytostabilization and preparing a contaminated site, before the decontamination by controlling the flow of contaminated groundwater and/or by determination of contaminated ingredients in the problematic area. This method is prospective, mainly for the remediation of contaminated soils, industrial waste landfills and waste piles. Limitation of the effectiveness of phytoremediation processes is in particular climatic conditions and local biogeographical factors (Mičieta and Polák 2000, Mičieta et al. 2002, Mičieta and Murín 2004).

The aim of our study is to show the practical outcomes of these theoretical guidelines at the three examples from Slovakia, situated in the very center of a Central Europe. Consequently these examples are useful for a larger part of Europe for the common flora and climate.

## 7.2. Materials and Methods

### 7.2.1. Areas of Remediation Implementing

#### *7.2.1.1. The Aluminium Plant*

It is located in the Žiarska kotlina basin in central Slovakia. The technological cycle includes not only the production of aluminium but also of $Al_2O_3$ and an electrode paste. Aluminium is obtained by electrolysis of $Al_2O_3$ with the use of Söderberg electrodes. Yearly, 600,000 t of bauxite and coke and 300,000t of brown coal (0.009% As and 29.55% of ash) was processed. More than 15,000t of emissions were produced per year, with 10,000 t being gaseous. Emissions of $F^-$ were 800-1000 t per year, the next is 42.5 t of As, 800 t of tar and phenols, 0.86 t of Hg and others (Maňkovská et al. 1988).

The waste dump is located on the West side of the complex of factory that was built in 1959, without a proper isolation and used till 1996. The total area is 40 ha, the height above the surrounding terrain is 42-45 meters, sometimes even more. The waste – red and brown sludge – was transported to the muddy fields at the top of the heap with water. An underground sealing wall around the whole perimeter was built in the years 1992-1997 to prevent penetration of contaminated water from the dump into the environment. The consequence is a gradual increase in the total accumulated amount of rainfall water in the dumping area. The overall average increase of water in the closed area is estimated up to 2.5368 $l.s^{-1}$ annually.

#### *7.2.1.2. The Nickel Plant*

The Sereď city is situated 62 km North-East of the capital city Bratislava. From the phytogeographical point of view the study area belongs to the Podunajská nížina Lowland (126 m above sea level), and it is located in the close vicinity of the river Váh. Production of nickel was based on ore imported from Albania. Its processing brought out great amounts of smelting waste and consequently the bulky heaps. Extension and height of these heaps interfere with the natural character of the landscape; they cause a constant deterioration of the ground waters and air in the vicinity of the Sereď city. Nickel production in Sereď was over during 1993.

#### *7.2.1.3. Landfill Pond of Waste Produced from a Generator Gas*

The repository waste site is located close to the village Kalinovo in Lučenec basin. Kalinovo is a small village which is 10 km North-East from the city Lučenec in south Slovakia, Lučenec is approx. 10 km from Hungarian borders towards the South direction (Figure 33).

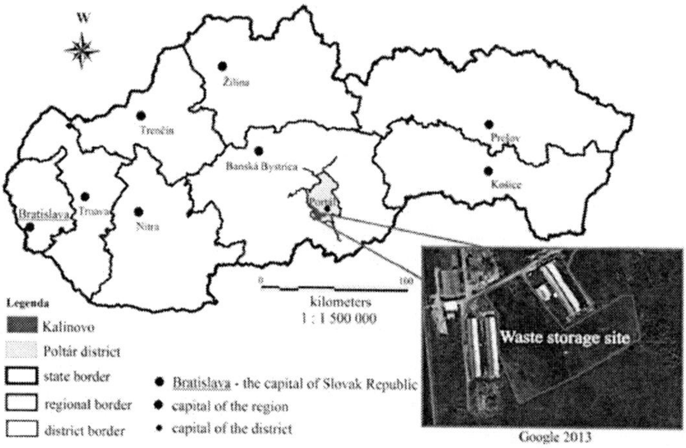


Figure 33. Landfill pond of wastein Kalinovo.

## 7.2.2. Level of Pollution

The composition and amounts of emissions from Aluminium Factory were so extensive that a presumption of not only toxic, but also mutagenic effects on living organisms is reasonable (more in Mičieta and Murín 2014). The territory within the radius of 10-15 km from the plant was visibly threatened. The location of the factory is in a basin, with a frequent occurrence of atmospheric inversions, which increase the concentrations of emissions containing fluorides, $SO_2$, As, PCB and PAH. Contents of metals as Al, As and Cr per 1 kg of soil were 127 – 156 mg, 40 – 160 mg and 60 – 160 mg, respectively (Holub, in litteris).

Negative effects on vegetation in the vicinity of a Nickel Factory were reported by Zeleňáková and Holub (1985). Soil around the heaps is polluted by the solid flue dusts of smelting waste containing heavy metals like Ni, Cr and Co. The accessible content of these elements is decisive for the plants. Extractible parts of smelting waste per 1 kg of soil were: Ni = 631 mg, Cr = 128 mg, Co = 185 mg (Zeleňáková and Holub 1985). Washed corn leaves collected in the vicinity contained about 10-fold and in the distance of 4 000 m from the dump double amounts of Ni, Cr, and Co, respectively, as compared to the control (Zeleňáková and Holub 1985).

Analysis at the storage site of waste produced from a generator gas confirmed the presence of hazardous chemicals (e.g., NEL = non-polar extractible agents per 1 kg of soil: 216 200 mg, phenol index: 732 mg). Observed qualitative indicators (from spectroscopy in the UV and IR level) of contaminated soil showed results only in the case of extractible and non-polar extractible agents extractives. An increase was also recorded in the case of phenols, that is corresponding to the chemical composition and nature of the waste land. Storage of waste is containing residues of gas tar and tar sludge generated as a by-product of the gas production which is evident with an increase in the values of NEL by spectroscopy in the UV level, which detects a conjugated unsaturated bonds compared to NEL with saturated bonds. This indicates that it is an organic pollution, mainly constituted by aromatic hydrocarbons. The analysis confirmed the presence of chemical compounds with hazardous characteristics (total aliphatic and aromatic hydrocarbons, phenols, polycyclic aromatic hydrocarbons). Its consistency is asphalt-like, solid at temperatures below 10°C before softening and liquefying at higher temperatures, emitting its characteristic smell. The total volume of waste mixed with soil is estimated to be about 6 000 $m^3$, which is confirmed by the results of geophysical and drilling works (area of about 1 200 $m^2$ with depth of about 5 m). Analysis of the extracts from waste in the year 1994 showed the presence of heavy metals too.

## 7.3. Results and Discussion

### 7.3.1. The Aluminium Factory

#### *7.3.1.1. Top of a Landfill Pond*

After a partial stabilization of the area free air raids of other species (such as *Anchusa officinalis, Artemisia vulgaris, Cirsium vulgare, Descurainia sophia, Conyza canadensis, Echium vulgare, Elytrigia repens*,

*Tanacetum vulgare, Daucus carota* and others) is expected with a perspective of the gradual establishment of a stabile subtermophil community, formed mainly by thermophytes and xero-thermophytes. The community, formed and stabilized this way, is in its development largely determined by the depth of the root system, but with a sufficient moisture we may assume creation of thermophilic ruderal vegetation outside the settlements. An important modifying factor will be a fertilization in subsequent years implemented at least 1x per year.

Due to a bentonite cover of the landfill pond, e.g., its isolation against rain, we proposed to use at least 10 cm soil layer with fixation of jute or sparse coconut network. Environmental and biological conditions require 10 cm of soil substrate. We assumed an initial application of granulated fertilizer at a rate of 1,000 g per 10 $m^2$. An area prepared this way is ready for implementation of grass mixtures for extremely dry locations and alkaline soils mixed with species for warm slopes like: *Festuca ovina*, *Lolium perenne*, *Festuca rubra*, *Poa pratensis*, *Agrostis tenuis*, *Agrostis stolonifera* with species of the family *Fabaceae* especially *Medicago sativa* or *M. lupulina*, *Onobrychys viciaefolia*, *Melilotus albus* and *M. officinalis*. Creating this grass community envisages the creation of a well rooted network underground capable of withstanding various irregularities and long periods of drought. Seed mixture is assumed to be 5 kg for 100 $m^2$.

#### *7.3.1.2. The Slopes of the Pond Achieving More Than 30 Degrees*

In these extreme places the first step is to lay the isolation layer of bentonite plus a double layer of a coconut mat with crushed straw, 10 cm layer is required. We proposed the implementation of seeds sowing with a sufficient drenched and fertilized mat with essential enrichment of humus component. For the seeds, it is appropriate to use a technical mixture for extremely dry areas like: *Festuca ovina andF. rubra*, *Lolium perenne*, *Poa pratensis*, even *Trifolium repens*, *Medicago lupulina*, *Securigera cook* and *Lotus corniculatus*, and if the second layer is realized, then *Onobrychys viciefolia* is also recommended. With a good rooting this vegetation will sufficiently resist to frost and drought. Extremity of slope requires permanent attachment network against water erosion by flushing the surface. An extreme slope and a thin layer of the substrate with root system will require cyclic irrigation and fertilization.

### *7.3.1.3. Terraces of Landfill Pond, of a Width of 2-5 Meters, Partly Filled up with Material from Upstream Slopes*

On the terraces we recommended at least 20 cm of soil required for a perspective growth of trees, with a base of herbal mixture as in the previous case. From introduced trees we recommended those well tolerant, growing and surviving plants, already verified on-site of pond Žiar nad Hronom, to be sowed in groups of each species: *Populus canescens*, *Salix caprea*, *Cerasus mahaleb*, also *Pinus nigra*, *Pinus sylvestris* and where appropriate *Betula pendula* and *Robinia pseudoacacia*. Due to the character of these terraces where they are supposed to survive, we may expect their inhibited growth and almost a shrub habit. Nevertheless, it is recommended to extend also the range of shrubs planted especially *Crategus* sp. div, *Prunus spinosa*, or *Eleagnus angustifolia*, *Lycium barbarum*, which is effectively strengthening the coherence of the soil substrate. With the given width of terraces we recommended to plant groups of chosen species in the clamps 2x2 m, e.g., at a width of 2 m only one row at a width of 5 m two rows.

### *7.3.1.4. Surroundings of a Landfill Pond, Which Is in South-West Part Waterlogged and Characterized by High Alkalinity Environment*

In this particular part of the pond, it is necessary to apply a mixture of grass species tolerant to the alkaline environment: *Festuca ovina*, *Lolium perenne*, *Festuca rubra*, *Poa pratensis*, also *Agrostis tenuis*, *Agrostis stolonifera*, with species of the family *Fabaceae*, especially *Medicago sativa* or *M.lupulina* and *Melilotus albus* or *M. officinalis*. In waterlogged sites, we recommended particularly species like *Populus canescens* and *P. nigra* sp. coll., and from the tolerating willows especially *S. alba*, *S. viminalis*, *S. caprea*, *S. cinerea*, but also *Populus tremula*, and for drier sites *Robinia pseudoacacia*, *Betula pendula* and *Acer campestre*. Planting buckles recommend are 2x2m. For better evapotranspiration of waterlogged areas we welcomed and supported spontaneous air raids of suitable and tolerant local species like: *Calamagrostis pseudophragmites*, *Phalaroides arundinacea*, *Agrostis stolonifera* and *Phragmites australis*.

### 7.3.2. The Nickel Factory

The special character of this site is caused by the close range to the river Váh, leading to the leaking of underground water within 600 m in capacity of 25 $l.s^{-1}$.

Phytostabilization of this dump is under the condition that at least 10-20 $l.s^{-1}$of this water will be drilled, resulting in a depression of the level of the underground water and thereafter avoidance of its contamination with the substrate. The litological situation is following: within 1.5 m = earth, till 2.5 m = sand and sand mixed with earth, till 8 m sandy gravel. Underground water is 2.0 – 2.5 m deep. Sanation of this dump by purely technical method will be unacceptably expensive. On the other hand, phytoremediation is offering 20 times cheaper (Chappel 1997) and ecologically acceptable solution (Dietz and Schnoor 2001). Fast-growing arborescent species are not only helping to drill troubling underground water, but also decreasing nitrogen agents in it five times (Clothier and Green 1997). Preferable specie in this case is *Populus canescens* with the ability to grow 3-5 m per year, high transpiration up to 100 l daily in age 5 years and high tolerance to organic contaminants. There is also a large scale of its cultivars available to use. We recommend sowing into two lines – the first right behind the dump and the second in a distance that is calculated by the stream of the underground water reached within a day, multiplied by the number of days of vegetation rest, approx. 200 down the stream. Lines will have a drainage function. Each line should be 5.5 m wide and 600 m long, covering altogether 1.3 ha. Bear in mind the position of the underground water; sowing has to be done 2 m deep. Every line has to consist of two rows with the sowing in an area of 2x2 m. As pioneering trees we recommended *Salix caprea*, which will open a space for *Populus canescens* within few first years. Therefore, the sowing has to be made by changing these two species one by one.

### 7.3.3. Landfill Pond of Waste from a Produced Generator Gas

#### *7.3.3.1. Top of a Landfill Pond*

Due to the already used folia isolation against the rain, we proposed to use at least 10 cm soil substrate. An area prepared this way is ready for implementation of grass mixtures for extremely dry locations and alkaline soils, mixed with species for warm slopes like: *Festuca ovina* and *F. rubra*, *Lolium perenne*, *Poa pratensis*, *Agrostis tenuis* and *A. stolonifera* with species of the family *Fabaceae* especially *Medicago sativa* or *M. lupulina*, *Onobrychys viciaefolia*, *Melilotus albus* and *M. officinalis*. Creating this grass community envisages the creation of a well rooted underground network capable of withstanding various irregular and long periods of drought. Seed mixture is assumed to be 5 kg for 100 $m^2$.

#### *7.3.3.2. The Slopes, Boundaries of the Pond*

For an establishment of a sustainable vegetation green cover, 10 cm layer is required. Before the seeds sowing sufficient drenching and fertilization is needed with essential enrichment of humus component. Seeds in the technical mixture are chosen for extremely dry areas, such as: *Festuca ovina* and *F. rubra*, *Lolium perenne*, *Poa pratensis*, even *Trifolium repens, Medicago lupulina*, *Securigera cook* and *Lotus corniculatus*. If the second layer is realized, then also *Onobrychys viciefolia* is recommended. With a good rooting this vegetation will sufficiently resist to frost and drought. Special attention needs to be focused towards the isolation folia that should not be penetrated by rhizosphere of later introduced trees.

#### *7.3.3.3. Terraces of a Landfill Pond Partly Filled Up with Material from Upstream Slopes*

On the terraces we recommended at least 20 cm of soil required for a perspective growth of trees, with a base of herbal mixture as in the previous case. From the introduced trees we recommended those well tolerant, growing and surviving plants, to be sow in groups of each species: *Populus canescens*, *Salix alba* and *S. fragilis* and *S. caprea*, *Cerasus mahaleb*, also *Pinus nigra*, *Pinus sylvestris* and where appropriate *Betula pendula* and *Robinia pseudoacacia*. Due to the toxicity of the basic substrate we may expect their inhibited growth and almost a shrub habit. Nevertheless, it is recommended to extend also the range of shrubs planted especially *Crategus* sp. div, *Prunus spinosa*, or *Eleagnus angustifolia* and particularly *Lycium barbarum*, which is effectively strengthening the coherence of the soil substrate. This also protects the remediation set from the invasion of weeds.

#### *7.3.3.4. Surroundings of a Landfill Pond*

In this part around a landfill pond we recommended to apply a mixture of grass species as follows: *Festuca ovina* and *F. rubra*, *Lolium perenne*, *Poa pratensis*, also *Agrostis tenuis* and *A. stolonifera*, with the species of the family *Fabaceae*, especially *Medicago sativa* or *M.lupulina* and *Melilotus albus* or *M. officinalis*. In waterlogged sites, we particularly recommended species like *Populus canescens* and *P. nigra* sp. coll., and from the tolerating willows especially *Salix alba*, *S. viminalis*, *S. caprea*, *S. cinerea*, but also *Populus tremula*, and for drier sites *Robinia pseudoacacia*, *Betula pendula* and *Acer campestre*. Planting buckles recommended are 2 x2m. For better evapotranspiration of waterlogged areas we expect spontaneous air raids of suitable and tolerant local species like: *Calamagrostis pseudophragmites*, *Phalaroides arundinacea*, *Agrostis stolonifera* and *Phragmites australis*.

## Conclusion and Future Prospects

It is common knowledge that phytoremediation is the most efficient in large areas where contamination reaches a depth of 5 m. All phyto-remediation processes are based mainly on strategies of phytostabilization and phytodecontamination (Glass 1999, Mičieta and Polák 2000 + 2001).

Vascular plants are an integral part of the natural environment and represent an important integrating sphere of the individual components of the environment, but a summary of information about their metabolism and degradation of xenobiotics is inadequate and the research still represents an initiation phase (Cunnigham and Berti 1993, Dietz and Schnoor 2001, Cunnigham et al. 1995).

The aim of the basic research is to find the most suitable plant species for phytoremediation (Schnoor 2000) under the following conditions: they have to be able to grow rapidly in the contaminated environment, to be resistant to contaminants like heavy metals and accumulated toxic metals by either transport into their peripheral organs, or metabolized in the non-toxic form. One of the most effective and best phytoremediation systems for implantation is the usage of particularly inexpensive or fast-growing pioneering trees.

Obviously local flora is the first on hand to be used although it is limited if selected species grow too slowly and do not produce sufficient amount of biomass. Another problem is if they are not present on the exposed locations or their reproduction in sufficient quantity is complicated. Therefore, projecting a particular phytoremediation at the proposed dump or landfill pond requires from its authors to be well oriented in the basic research of systematic botany, enabling the selection of the appropriate range of species, in extreme cases the spectrum of adapted and tolerant genotypes of appropriate species or their cultivars.

A limiting factor in the use of plants is a long time for decontamination. That is why for the near future is our aim to obtain special plants adapted to the requirements of phytoremediation by breeding and genetic engineering (Frančová et al. 2001).

*Chapter 8*

# RETROSPECTIVE BIOMONOTORING: A HUNDRED YEARS OF ENVIRONMENTAL POLLUTION AT SELECTED AREAS IN SLOVAKIA

## 8.1. INTRODUCTION

With increasing demands for the protection of our environment, flowering plants have been used as bioindicators of mutagenicity, phytotoxicity and genotoxicity of environmental chemicals (Constantin and Owens 1982, Grant and Zura 1982). A WHO/IPCS collaborative study on plant test systems was focused on standardization of certificated tests for genotoxic effects of xenobiotics (*Allium*, *Vicia*, *Arabidopsis* and *Tradescantia*), and their application to "*in vivo*" conditions established (De Serres 1992, Kanaya et al. 1994). The methodology of the plant tests with the above-mentioned species is now a routine (De Serres 1978, De Serres 1992, Mičieta 1990). However, the native flora and crop plant species in "*in situ*" conditions also appear to be capable of serving as bioindicators of genotoxicity (Sharma and Raju 1983, Tomkins and Grant 1976, Mičieta and Murín 1996a). The main advantage of using natural flora for the assay is a ready availability and variety of assay material, offering different periods for sampling. Assays can be undertaken employing flora from test and control areas.

The process of "*in situ*" indication of genotoxicity requires the observation of standardized methodological procedures and a well-founded selection of indicator species in the broadest possible spectrum to obtain a reliable indicator of the genotoxic potential in the examined locality or region. Prior to the present study there have been no reports aimed at establishing a set of standard, reliable bioindicators from wild plant species; there were reports involving plants generally (De Serres 1978 + 1992, Mičieta 1990, Mičieta and Murín 1996a), or particular species (CelaRenzoni 1990, Mičieta and Murín 1996a, Murín 1987 + 1995).

In this chapter we would like to compare the obtained results of current environmental pollution at different areas in Slovakia with herbarium specimens which were collected at the same areas within more than one hundred years. Luckily these specimens are still preserving the effects of pollution at the time of collection and are suitable for retrospective evaluation of pollution. Our aim has been to locate the "point zero" in the areas we chose for our study; in other words, to pinpoint the moment when pollution had a perceptible effect on the environment of these areas because of their industrialization for the first time, and to start our investigation from that point.

## 8.2. Material and Methods

Apart from usual sampling, our original method of the retrospective biomonitoring of the environmental pollution, invented by prof. Karol Mičieta, was used. Thanks to this method we are able to test the genotoxic pollution of the environment up to 100 years back (Mičieta and Murín 1999). It allows us to return to the "point zero," when there was no industrial activity in the chosen region, and compare it with the current situation. We may also observe the whole history of genetoxic effect of pollution in this region between these two data-points. Retrospective biomonitoring was made possible by the collection of herbaria specimen of

Department of Botany Comenius University in Bratislava headed by prof. Karol Mičieta. These contain pollen samples from chosen areas in Slovakia since the end of the 19th century. The comparison of the tetrade analysis of pollen of *Calluna vulgaris* from the latest specimens with those collected back in the years 1944, 1941, 1926 and 1894 showed an interesting dynamic of pollution impact.

*Calluna vulgaris* was chosen not only because it is widespread in Slovakia, and thus commonly found in both polluted and non-polluted areas, but because of its persistent pollen tetrads. These attributes make it easier to distinguish physiological and genetic damage to the pollen grains. The abortion of one, two or three pollen grains in a pollen tetrad usually indicates genetic damage. Physiological damage usually affects the whole flowers, the anthers, and each of the four pollen grains in the pollen tetrads. Genetic damage of all four pollen grains, however, is so rare that although we might exclude such data mistakenly as physiological damage, it has no statistical significance (Mulcahy 1981).

Old pollen grains from herbaria samples were stained with 0.5% aceto-carmine or 0.05% aniline blue in lactophenol (Kearns and Inouye1993) adding extra 60 min for slow warming up to 60 °C in distilled water for their better stainability.

The tested area is described in chapter 3.

## 8.3. RESULTS

With the methods described above, we were able to establish an ecological history for the Žiar nad Hronom area. Our research covered a period starting from the year 1894 and ending in the year 1998. Old pollen grains from the polluted areas, in comparison with equally old pollen grains from the control areas, showed significant differences in pollen abortions in relation to the place and the age of the specimens collected (see Table 42). Samples of tetrade analyzed *Calluna vulgaris* during the last one hundred years were tested in three regions of Slovakia: the heavily polluted area of Žiar nad Hronom, and two control areas (the Záhorská nížina Lowland and the Vysoké Tatry -- "High Tatras" -- mountains). It is interesting to note that the highest rate of aborted pollen tetrads (up to 56% of aberrant pollen tetrads!) among specimens collected at Žiar nad Hronom seems to have occurred in the year 1965. In that year, the Aluminium factory's production reached its peak, while measures undertaken to contain the emissions at the factory were minimal. The subsequent decline of the effect of pollutants on tested pollen grains corresponds with the historical trend of decreasing production and simultaneously increasing environmental awareness (especially after the year 1989) with the latest samples (15-21%). These figures (see Table 42) may be the result of the decrease of the amount of emissions due to the current environmental policies, the decrease in overall production as well (especially after the year 1989), and the possible adaptation or tolerance of the exposed plant population (Mičieta and Murín 1998a + 2014). Surprisingly different results were obtained from the control areas. While the first control area (the Záhorská nížina Lowland) confirmed our expectations of minimal pollution, we observed surprisingly high rates of abortiveness in the High Tatras mountains (average altitude 2,000 m), un-industrialized and popular recreational area. It is also used as a climatic spa (sic!). The content of heavy metals in pollen and seeds of Norway spruce (*Picea abies* (L.) Karst.) had already been tested there (Ostrolucká and Fleischer 2003). A possible explanation is that this area is contaminated by long-rank pollution from the heavy industrial areas around the Polish coal-basin (Ciezsyn, Katowice) approx. 200 kilometers to the North-West.

Importance of this example of the highest polluted site in Slovakia, Žiar nad Hronom, is leading to the latest study of Pogányová et al. (2017) with a following data of general abortiveness in % in the following years: 2010 = 5%, 2011 = 6.2%, 2012 = 4.9% and 2013 = 5.3%. These data are confirming a decrease of the impact of the formerly critical pollution of the site by heavy metals.

However, thanks to our herbaria specimen we may trace the situation of the air-pollution in Bratislava city centre too, in this case back to the year 1866. Here we may see a significant increase of pollution in 1962, repeated again in 1987 and after (Figure 34).

## Table 42. Tetrade analyses of *Calluna vulgaris* (L.) Hill

| Area | Collector | Year | Abortiveness | Abortiveness | Abortiveness |
|---|---|---|---|---|---|
| | | | Genetic in % | Physiologic in % | General in % |
| Záhorie | | | | | |
| /lowlands/ | Ptačovský | 1926 | 2.7 ± 0.3 | 0.4 ± 0.1 | 3.1 ± 0.3 |
| | Futák | 1941 | 2.4 ± 0.5 | 0.3 ± 0.1 | 2.7 ± 0.4 |
| | Mičieta | 1991 | 5.1 ± 0.2 | 0.2 ± 0.1 | 5.3 ± 0.3 |
| | Mičieta | 1995 | 5.0 ± 0.7 | 0.4 ± 0.1 | 5.4 ± 0.6 |
| | Mičieta | 1998 | 5.0 ± 0.4 | 0.2 ± 0.1 | 5.2 ± 0.3 |
| Vysoké Tatry | | | | | |
| /mountains/ | Greschik | 1894 | 2.4 ± 0.6 | 0.3 ± 0.1 | 2.7 ± 0.6 |
| | Futák | 1944 | 8.4 ± 0.9 | 0.5 ± 0.2 | 8.9 ± 1.0 |
| | Mičieta | 1991 | 24.4 ± 1.5 | 2.0 ± 0.3 | 26.4 ± 1.7 |
| | Mičieta | 1995 | 18.9 ± 1.8 | 1.9 ± 0.5 | 21.8 ± 1.9 |
| | Ďurišová | 1998 | 20.6 ± 1.4 | 1.4 ± 0.2 | 22.0 ± 1.4 |
| Žiar n. Hronom | | | | | |
| /alum. plant/ | Kmeť | 1894 | 4.9 ± 0.5 | 0.8 ± 0.2 | 5.7 ± 0.7 |
| | Hajdúk | 1965 | 52.0 ± 2.8 | 4.7 ± 0.7 | 56.7 ± 2.6 |
| | Mičieta | 1991 | 19.9 ± 0.8 | 1.6 ± 0.8 | 21.5 ± 1.1 |
| | Mičieta | 1995 | 17.4 ± 1.7 | 1.5 ± 0.4 | 18.9 ± 1.4 |
| | Mičieta | 1998 | 14.5 ± 1.2 | 1.2 ± 0.2 | 15.7 ± 1.4 |

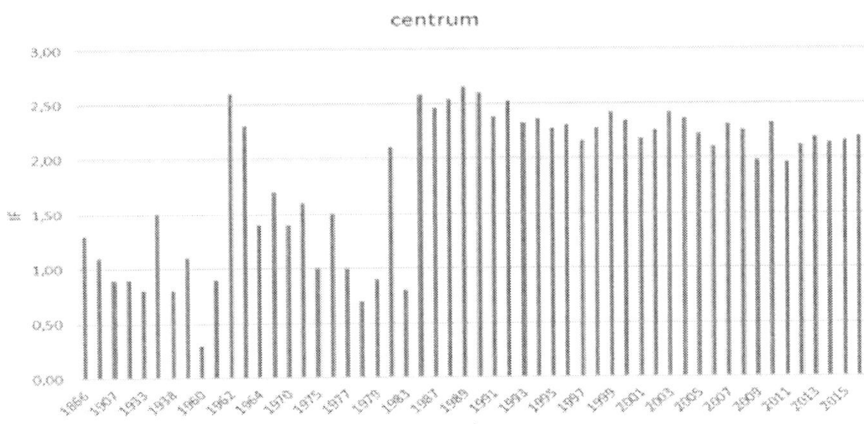


Figure 34. Long-term biomonitoring of the environment of Bratislava city center using herbaria specimen and samples.

## Conclusion and Discussion

An increased level of five PAH in the samples of *Calluna vulgaris* (Mičieta et al. 1998) lead us to the conclusion that the bioindicator used in this study is able to sensitively detect the presence of the above mentioned pollutants in a chosen area and is utilizable at other localities too.

There are several advantages in using microspore mother cells, microspore tetrads and pollen grains as bioindicators of mutagenicity. One is the high sensitivity conditioned by the haploid state, when all lethal mutations affecting the development of pollen are evidenced immediately. The large quantity of pollen allows a sufficient set of evaluated individuals for the quantification of the results. In an evaluation of the pollen abortivity an exceptional sensitivity, when deletions in virtually any part of the genome will induce pollen abortion, must be taken into consideration (Mulcahy1981). However, the greatest advantage is that the species live directly in the polluted environment under evaluation, and this makes it possible to evaluate the effectiveness of ecological factors including pollution components. The method is efficient and easily applicable. It represents a simple and economical procedure that is suitable especially for long-term observations.

However, according to our knowledge, it is for the first time that a plant test is allowing us to check the environmental situation in a particular area retrospectively. We hope that this method invented by prof. Karol Mičieta will give us a chance to reconstruct the history of areas of our interest, with a special emphasize to the "point zero" of environmental pollution, which was developed here during the industrial and postindustrial period.

*Chapter 9*

# PARALLEL PLANT TEST OF PHYTOTOXICITY AND MUTAGENICITY PRESENTED ON PRACTICAL EXAMPLES OF POLLUTED SITES IN SLOVAKIA, CENTRAL EUROPE

## 9.1. INTRODUCTION

Our original method of "in situ" biomonitoring of a polluted environment by wild plant and crop plant species, which allows us to use it even for a retrospective monitoring (see Chapt. VIII), is extended here by another original test invented in our laboratory. Practical examples of its usage are included.

Tests of the toxicity and genotoxicity of different xenobiotics has great practical importance, but their detection and characterization only by chemical analysis is limited, and the interpretation of their impact on the biota is difficult (Speranza et al. 2010; Kuchma and Finkeldey 2011). Different interactions between individual components and their bio-metabolism occur in every ecosystem, e.g., in urban agglomerations (Bodini 2012). It is difficult to determine only from the knowledge of its individual constituents how a certain mixture of different xenobiotics will interact (Claxton and Woodall 2007). Suitable "*in situ*" tests by means of bioindicators are required to detect the xenobiotics effects in our environment. Plants have an unsubstituable position in this process as sensitive and effective bioindicators (Grant 1998). Their genetic system provides excellent "*in situ*" environmental monitoring (Käffer et al. 2012). Their advantages include easy manipulation and ability to metabolize non-mutagenic pollutants to mutagenic transmetabolites with an influence in the food chain too (Lakes and Kim 2012).

Toxicity plant tests recommended by legislation of The Organization for Economic Co-operation and Development (OECD 1984, 2000) and United States Environmental Protection Agency (US EPA 1985, 1996) include: growth test, root growth test, life cycle test and fluoromethric analysis of photosynthesis. International cooperation in standardization of tests for mutagenesis under International Programme on Chemical Safety (IPCS), World Health Organization (WHO), United Nations Environment Programme (UNEP) and International Labour Organization (ILO) has introduced test protocols for *Tradescantia*, *Vicia faba* and *Arabidopsis* (Kanaya et al. 1994, Grant 1998). Following this basic protocols we had used a parallel plant test where the test of phytotoxicity is enlarged by the parallel test of mutagenicity. In this chapter we are presenting practical examples of its use on the environmentally polluted sites in Slovakia, Central Europe. This report is based on more than two decades of our experiences with the detection of phytotoxicity and mutagenicity by means of wild plant species. In the presented method the usual phytotoxicity parameter of environmental pollution is extended by a parallel evaluation of mutagenicity in the same plant's bioindicator species. Practical examples are shown on the waste storage site for residuals from the generator gas at Kalinovo village and the Ore Mine Area. Special attention was paid to the relationship between chromosome aberrations and the nuclear DNA content, the number of mitotic cycles and the rate of growth inhibition. Knowledge of these relationships enables a correct interpretation of the test results. The parallel plant test is practical method for the risk assessment of xenobiotics in the environment.

The presented test is suitable for the evaluation of the effects caused by xenobiotics in both liquid and solid medium.

## 9.2. Materials and Methods

### 9.2.1. Sampling Site

Kalinovo is a small village 10 km North-East from the city Lučenec at the southern borders of Slovakia. Lučenec is approx. 10 km to the South from Hungarian borders (Figure 35).

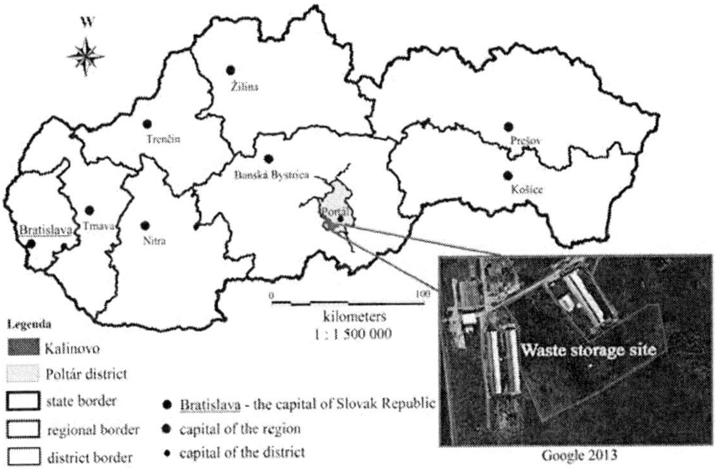


Figure 35. Localization of the tested site: generator gas waste storage site at Kalinovo.

The Žiaromat generator gas waste storage site, located close to the village Kalinovo was chosen as an example of a heavily polluted industrial complex. The waste at this site contains tar and sediment residues from the gas production. The total volume of waste mixed with soil is estimated to be about 6,000 $m^3$, which is confirmed by the results of geophysical and drilling works – an area of about 1,200 $m^2$ with a depth of about 5 m (Sámelová et al. 2011). Analyzes of the waste from the production of generator gas at the storage site confirmed the presence of hazardous chemicals (e.g., NEL = non-polar extractible agents: 200 – 216 $mg.kg^{-1}$, phenol index: 732 $mg.kg^{-1}$). Observed qualitative indicators (from the spectroscopy in the UV and IR level) of contaminated soil showed results only in the case of extractible and non-polar extractible agents (Sámelová et al. 2011). Storage of waste contains residues of gas tar and tar sludge generated as a by-product of the gas production with an evident increase in the values of NEL confirmed by spectroscopy at the UV level. This method detects conjugated unsaturated bonds in comparison to NEL with saturated bonds, indicating organic pollution constituted mainly by aromatic hydrocarbons. The results of chemical analysis confirmed the presence of chemical compounds with hazardous characteristics as total aliphatic and aromatic hydrocarbons, phenols and polycyclic aromatic hydrocarbons. Its consistency is asphalt-like, solid at temperatures below 10°C. When softening and liquefying at higher temperatures, it is emitting a characteristic smell. Analyzes of the extracts from the waste also showed the presence of heavy metals.

### 9.2.2. Selection of Plant Bioindicators and Their Cultivation for Testing

Several bioindicator plant species, like *Raphanus sativus* L. *subsp. sativus var. radicula, Vicia sativa* L., *Vicia faba* L. and *Zea mays* L., were studied in our previous experiments for the toxicity detection. These can be used in laboratory conditions for precisely defined chemical concentrations under controlled application, exposition time, temperature and nutrition conditions. The method used herein is based on the protocol of Murín (1984), observing the seed germinability and seedling growth after 72 and 96 h, at 25°C temperature. Placed in the vertical cultivation container it presents optimal conditions for mitotic and chromosome cycles (Murín 1987). The simple growth test (OECD, 2000) recommends the use of three bioindication groups: *Brassicaceae* and *Viciaceae* (among *Dicotyledones*) and *Poaceae* (among *Monocotyledones*).

For a parallel testing of physiological and genetic toxicity we recommend the usage of one- or two years old seeds imbibed for 24 h before being treated by test samples. Chosen seeds are germinated in large Petri dishes (preferably with a diameter of 23.5 cm and a capacity of approximately 2,000 $mm^3$). We recommend 25 (5x5) or 36 (6x6) imbibed seeds per one Petri dish. Numbers of seeds are important when we consider a consumption of oxygen by germinating seeds. An inadequate large number of seeds per Petri dish is causing a hypoxy and thus confusing results especially in terms of mutagenicity (Sámelová et al. 2011).

### 9.2.3. Phytotoxicity Test

This test required different conditions for cultivation of plant species generally and *Vicia faba* L. especially.

#### *9.2.3.1. Basic Procedure*

Chosen seeds were imbibed for 24 h in dilutions by distillated water of wastewater or water extract from the soil taken directly from the environment. Materials from the waste site were transported to the laboratory in sealed and marked containers. The recommended dilutions of the wastewater or water extract from the soil are in multiples of 10 first, ranging from 1:10 to 1:10.000, and then more precisely in an effective scale of the environmental impact found. In case of different studied chemical agents, concentrations can range from 1 mg to 10 g in 1 l of distillated water.

The seeds were then placed on Petri dishes with three layers of filter paper soaked to saturation with the same dilution or concentration as the tested samples. Cultivation containers can be used depending of a size of seeds and they need oxygen while germinating. The Petri dishes or closed vertical cultivation containers were placed in the thermostat at 25°C temperature for cultivation of seeds for 72 h. The roots were measured to determine the level of growth inhibition and any abnormalities. The control seeds were imbibed in the distillated water.

#### *9.2.3.2. Vicia faba Seeds Modification*

The *V. faba* seeds required an individual treatment for this test especially because of their consumption of oxygen when germinating. It is recommended to remove seed coat or at least penetrate it by lancet in case of imbibed seeds. Imbibition in the distillated water should last for 6–18 h, and then the imbibed seeds should germinate in vertical cultivation containers filled with wet sawdust or perlite at 25°C for 60–64 h. Seedlings with 30–40 mm long roots have to be washed and then used for testing particular samples of toxic compounds. For each sample 10 seedlings are placed on perforated lids, with their roots soaked in the tested dilution, and left to grow for 24 h, before being transferred to distillated water for the following 24 h. Then, the primary roots are measured. When a significant growth inhibition is indicated all roots are fixed for mutagenic testing.

### 9.2.4. Contact Soil Test

In this test the seeds are planted directly into the tested soil. The imbibition of seeds in the distillated water lasts only for 2 h, maximum of 8 h. For the dilution of tested soil samples it is recommended to mix them with a neutral soil consisting of at 1/3 of sea sand, 1/3 of peat and 1/3 of kaolin. The final mix is watered by the distillated water. The duration of the test is 96 h. Seedlings are then carefully cleaned from the tested soil sample and the test continues under the standard procedure (Murín and Mičieta 2009b).

### 9.2.5. Test of Mutagenicity

After the cultivation and measurement in the phytotoxicity test, roots with growth the inhibition of 25, 50 and 75% were used to determine their mutagenicity. Average-length roots were taken from each sample and their tips were fixed in ethanol: acetic acid (3:1). The fixed root tip length is recommended to be approximately 3 mm. The samples left in the fixation liquid were evaluated, or after 24 h preserved in 75% ethanol for later processing.

The root tips were macerated in ethanol:36% hydrochloric acid (1:1) for 7–8 min, then washed with the distillated water for 10 min and used for squash slides stained with acetic-orcein or Feulgen reaction (Darlington and La Cour 1976).

These slides were examined for chromosome or genome mutations. For the cytotoxicity evaluation we chose ana-telophases, in accordance with other authors (Bezrukov and Lazarenko 2002). Approximately 300 ana-telophases were evaluated in the selected bioindicators from *Viciaceae (Vicia sativa* L., *Vicia faba* L.), *Brasicaceae (Raphanus sativus)* and *Poaceae (Zea mays* L.). Results were calculated by ANOVA analysis, where mutagenicity is established at the Poisson distribution, according to Amphelett and Delow (1984).

For easier comparison of results we recommend to use index of inhibitory concentration ($IC_{50}$), referring to the well-known formulas of lethal dose (e.g., $LD_{50}$) and lethal concentration (e.g., $LC_{50}$). $IC_{50}$ than means that evaluated pollutant is causing reduce of growth of plant roots to 50% of its normal growth as indicated by control.

## 9.3. Results

### 9.3.1. The Relationship between the Frequency of Chromosome Aberrations and Mitotic Cycles

#### *9.3.1.1. Example from Kalinovo Storage Site*

To adjust this test for a practical use we may take into account in what mitotic cycle can be done sampling and evaluation of an impact of selected pollutant to the plant cells. There is an extensive knowledge on the mitotic cycle and its consequence for both *V. faba* L. and *Vicia sativa* L. species (Murín 1987) and therefore the relationship with chromosome aberrations can be accurately expressed. Differences in the onset of the first and second mitosis can also be readily observed in these species. The first mitosis in *V. sativa* radicules occurs 36h after seeds placement in the water, but for *V. faba* this takes 48h. Our results show an increase in chromosome aberrations in the following mitotic cycles in long-term seeds and seedlings treatment (Figure 36).

The water extract from the generator gas waste showed significantly ecotoxic and mutagenic effect (see Table 43).

**Table 43. Simultaneous test of mutagenity and phytotoxicity of the waste from generator gas at the Kalinovo storage site**

| Tested sample | Root length (mm) | IC v % | Ana-telophases evaluated | Number of aberrations | Frequency of aberrations (%) |
|---|---|---|---|---|---|
| *Vicia sativa* | | | | | |
| $H_2O$ | 31.4 ± 2.1 | 00.0 | 500 | 1 | 00.2 |
| MH 0,1mM | 22.9 ± 1.8 | 27.1 | 500 | 121 | 24.2 |
| Unadulterated | 0 | 100 | - | - | - |
| ½ dilution | 12.5 ± 0.6 | 60.0 | 300 | 37 | 12.3 |
| ¼ dilution | 18.9 ± 0.9 | 40.0 | 300 | 27 | 09.0 |
| 1/8 dilution | 28.2 ± 1.1 | 10.0 | 300 | 10 | 03,3 |
| *Zea mays* | | | | | |
| $H_2O$ | 55.9 ± 3.3 | 00.0 | 500 | 1 | 00.2 |
| MH 0,1 mM | 48.3 ± 4,2 | 13.6 | 500 | 99 | 19.8 |
| Unadulterated | 0 | 100 | - | - | - |
| ½ dilution | 22.4 ± 2.6 | 61.0 | 250 | 25 | 10.0 |
| ¼ dilution | 36.9 ± 2.8 | 34.0 | 250 | 16 | 06.4 |
| 1/8 dilution | 55.2 ± 3.1 | 11.0 | 250 | 6 | 02.4 |
| *Raphanus sativus* | | | | | |
| $H_2O$ | 49.8 ± 2.9 | 00.0 | 500 | 1 | 00.2 |
| MH 0,1 mM | 36.5 ± 3.1 | 26.8 | 500 | 83 | 16.6 |
| Unadulterated | 0 | 100 | - | - | - |
| ½ dilution | 21.5 ± 1.9 | 56.8 | 300 | 33 | 11.0 |
| ¼ dilution | 31.8 ± 2.3 | 36.1 | 300 | 20 | 06.7 |
| 1/8 dilution | 45.6 ± 4.1 | 08.4 | 300 | 7 | 02.3 |

Legend: $H_2O$ = negative control, 0.1mM maleic hydrazide (MH) = positive control, IC = index of inhibitory concentration.

Despite the poor dissolubility by distillated water of polyaromatic compounds, 1/2 dilution of the waste samples still causes 50% phytotoxicity, and all dilutions cause significant chromosome mutagenesis.

These results are of utmost importance for appropriate future management of this waste site and its phyto-remediation.

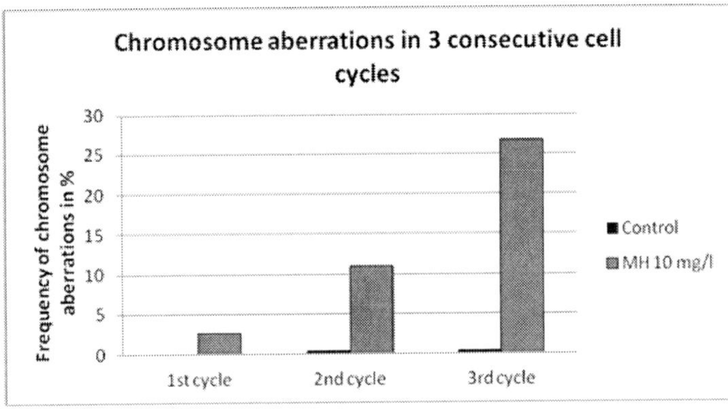


Figure 36. Frequency of chromosome aberrations during three mitotic cycles in the root tips of *V. sativa*.

### *9.3.1.2. Example from the Ore Mine Area*

The advantage of the Parallel test of phytotoxicity and mutagenicity is, above all, to compare the effects (of tested chemicals, polluted waters and others) on chromosomal structure and inhibition effects on reproduction cycle of testing plants (Murín 1984). Example is shown in tests from the Ore Mine Area (for a details see chapter 3).

Leach of sludge in the asserted solutions inhibited root growth of *Vicia sativa*, but the results of chromosomal aberrations showed significant differences only after the exposition to the maximum concentration. As a positive control was used a mutagen maleic hydrazide that caused a several times higher frequency of chromosomal aberrations against control. Interestingly first three dilutions of the leach of sludge had growing inhibitory effect, but dilution to 1:16 and 1:32 appear to be stimulating to the growth of roots of *Vicia sativa* L. seeds. Samples of the water from Pezinok brook appear to be a strong stimulant while the mining water was a strongest inhibitor of the growth of roots of *Vicia sativa*. Regardless of such a significant results in this parameter, differences of chromosomal aberrations in mitosis when compared with the results from the negative control are not so strong (Table 44).

Significant results were ascertained in the case of maximum ratio of sludge solution used in the "*in vivo*" conditions.

**Table 44. Parallel test of phytotoxicity and mutagenicity effects on *Vicia sativa* L., after the treatment of sludge solutions**

| Chemicals | Dose/Ratio | Phytotoxic effects (% of inhibition) | Aberrations (%) |
|---|---|---|---|
| Leach of sludge | 1: 2 | 57 | 2.67 ± 1.00* |
| | 1: 4 | 32 | 2.33 ± 1.33 |
| | 1: 8 | 21 | 1.66 ± 0.88 |
| | 1: 16 | stimulant | 0.33 ± 0.33 |
| | 1: 32 | stimulant | 0.00 ± 0.00 |
| Water of Pezinok brook | 1: 1 | strong stimulant | 0.00 ± 0.00 |
| Water of mine | 1: 1 | 9 | 0.00 ± 0.00 |
| Control (+) | 11.2 ug/ml | 70 | 10.00 ± 3.00* |
| Control (-) | | 0 | 0.00 ± 0.00 |

Control (+) = maleic hydrazide, Control (-) = distillated water, * significant at 5% level.

## 9.4. Discussion

We recommend the species *Sinapis alba* L., *Raphanus sativus* L. *subsp. sativus var. radicula, Vicia sativa* L. and *Zea mays* L., selected in agreement with valid norms for their easy usage and validity of the obtained results (Mičieta and Murín 2002b).

### 9.4.1. Evaluation and Interpretation of Phytotoxicity

The primary indicator of the toxic effect of the studied waste water, soil extracts and chemical samples is the change in seed germinability and in the growth of seedlings in controlled cultivation conditions. These changes are most often observed as growth termination, or more commonly, a slowing down the growth rate.

Growth is a complex process; including cell division, growth and differentiation, and each of these processes can be primarily or secondarily disrupted at any stage. Defects affecting these processes become evident in the first stages of the experiment through the change or delay in onset of germination. In the later stages, defects are evident in seedling growth rate. In conclusion, even if only two parameters – germinability and root growth – are monitored in the test of phytotoxicity, these parameters include all basic biological processes typical in plants (with the exception of photosynthesis).

Interpretation of toxicity is usually presented in $LD_{50}$ and $LC_{50}$. However, for evaluation and comparison of results from the short-term tests we recommend an inhibition index – $IC_{50}$. It defines the dilution of the sample, or concentration of the tested chemical, which inhibits seedling root growth to 50% of control inhibition. This toxicity measure indicates plant reaction sensitivity sufficiently adequate for $LD_{50}$ value comparison. The main reason for the choice of this toxicity measure is that in short-term plant toxicity tests it is not always possible to objectively determine whether a seed or seedling is dead, or just stagnant in growth. In addition, the $IC_{50}$ inhibition index expresses the extent of damage in the whole set of seedlings more accurately; including individual variability in sensitivity to the tested sample. The 50% growth inhibitions makes the vegetation period last twice longer than normally, and therefore affected plants are unable to complete their vegetation and reproduction cycle. Consequently $IC_{50}$ inhibition index may have a fatal impact on them.

The principle of the growth test in eco-phytotoxicity is the observation and measurement of roots in the seedlings of three species. Here, both *Sinapis alba* L. and *Vicia sativa* L. are dicotyledons, while *Zea mays* L. is a monocotyledon. Comparison of the toxicity of the tested samples involved inhibition concentration or dilution values of $IC_{100}$, $IC_{50}$ and $IC_0$. This indicated 100%, 50% and zero inhibition in the bio-indication species' seedling root growth under the defined conditions and compared to the control. For example, the phytotoxicity of waste water can be expressed as: $IC_{100}$ = 1:0, $IC_{50}$ = 1:20 and $IC_0$ = 1:10.000, where dilution of the tested sample with clean water in the ratio 1:10,000 has no observable toxic effects.

### 9.4.2. Evaluation of Chromosome Aberrations and Defects in the Mitotic Spindle

Evaluation of mutagenesis in the tested sample involved chromosome aberrations and genomic mutations caused by defects in the mitotic spindle.

Chromosome aberrations include chromosome fragmentation and bridges, and these are best evaluated in the late anaphase and the early telophase (ana-telophase) of the cell cycle. When they are significantly more abundant than the spontaneous mutability level, the types of chromosome aberration can be further specified in c-metaphase in a follow-up evaluation where the roots are treated with 0.05% colchicine 3 h before fixation (Gichner et al. 1994).

Mitoclasticity and genomic mutagenesis indicate defects in the mitotic spindle which result in an altered chromosome number. These defects can be observed in the same slides as chromosome aberrations, and they are divided into the following three categories:

1. Delayed chromosomes
2. Multipolar figures
3. Complete disintegration of the mitotic spindle

The three defect categories revealed in this test elucidate a potential danger to plants and eucaryotes. It is remarkable that these serious defects are not legislatively tended by toxicity tests in eucaryotes. Most importantly, the treatment of tested cells with a c-mitotic chemical and evaluation of chromosome and chromatid aberrations in c-metaphase make identification of these defects impossible.

### 9.4.3. The Relationship between Growth Inhibition and Mutagenicity

The advantage of this Parallel test is that it provides the opportunity to observe both growth inhibition and mutagenicity markers at the same time. The general rule is that higher toxicity should be accompanied by an increase in both growth inhibition and mutagenesis. The reliability of this rule, however, is challenged by changes in cells and mitotic cycles due to the influence of toxic chemicals which also result in changed frequency of aberrant chromosomes. Because the number of mitotic cycles decreases with stronger growth inhibition, it is necessary to correlate mutagenic data not only with the concentration of the sample, but also with the rate of growth inhibition and indirectly with the number of mitotic cycles. The most favorable correlation occurs at 50% growth inhibition. When mitotic activity at this inhibition rate is too low or zero, it is preferable in mutagenic analysis to choose plant roots with lower inhibition; for example 25%. On the other hand, defects in the mitotic spindle are more easily observable in samples with higher growth inhibition; up to 75%. Most importantly, the knowledge of growth inhibition rates identifies whether the studied chemical or compound permeates more complicated and costly tests based on mammalian or human cells.

## Conclusion

The methodology of the parallel plant test of phytotoxicity and mutagenicity enables a qualified detection of both toxic and mutagenic effects of a particular chemical compound, or their mixture in the real complex of the environment, which is important for an accurate risk assessment. This test is built upon the current legislative of recommended methods and it proved to be reliable with different chemicals and samples taken directly from the environment, as illustrated on the example of the

Kalinovo gas waste storage and the Ore Mine Area. Finally we may recommend it as an effective tool for testing phytotoxicity and genotoxicity "*in situ*."

*Chapter 10*

# CONCLUSION AND PERSPECTIVES

Environmental pollution is a global issue touching every society on the Earth. Our study is demonstrating it mainly on examples from Central and Eastern Europe. We might know that in the 1980's countries in this region adapted very strict rules to curb pollution. However, the law enforcement failed in many cases because of the utopic, un-natural character of such a society. Restriction and financial punishment of those "pollution producers" was rare in practice because of the general excuse that these "publicly owned" factories would fall short of their production quotas. And the production was believed to be a law itself pretending to make it good for everybody. In fact, the sustainable managed development of a healthy environment is unimaginable without respecting the natural mechanisms. Communism, demonstrating a certain good will, had ignored the fact of death and decay in the economic sphere. Non-effective companies had become "immortal" and consequently killed their patron, the communist regime. It had become unable to compete with other economic systems that respect this (at first look a negative) element of natural laws.

One of the self-saving mechanisms of Nature is an auto-regulation and we may also see it in purely human activities. In the seemingly antagonistic relationship of ecology and economy a new important progressive element has shown up in countries that adopted a free-market economy. Ecological thinking had become an important factor in the global market – so-called "environmental friendly products" are nowadays a strong weapon of competition. In the near future, disregarding environmental criteria will be fatal for companies. This surely is a welcomed auto-regulating element, the occurrence of which is to be supported in the interest of the sustainable managed environmental development. The "corporate greening" became the question of the summit of international leaders in the United Nations in June 1997 when they evaluated progress since the 1992 Earth Summit. However, they had to admit that government and institutional aid has fallen, leaving more responsibility for green innovations to the free market and more global initiatives for next decades up to these days.

## 10.1. Self-Saving Mechanism

Nature is resisting our interferences mainly due to the auto-regulating and repair mechanisms. Adaptation is another mechanism, thanks to which Nature shows its flexibility. Even environments so heavily damaged by human activities that they become highly risky for human survival; lower organisms such as e.g., extremophiles can still adapt (Ebringer, Krajčovič 1994). Our original "in situ" method of biomonitoring of polluted environment was tested in different field studies over the past thirty years. It showed its capability both for industrial and municipal areas. Among these tests, the particularly interesting were:

- long-term biomonitoring of the polluted environment in Kuwait after the Gulf war (Malallah et. al. 1997),
- retrospective biomonitoring of the environmental pollution in the vicinity of heavy industrial areas in Slovakia (cf. Murín and Mičieta 2000)

- bioindication of radioactive-contaminated areas around the Nuclear Power Plant Jaslovské Bohunice (Mičieta and Murín 2007).

Our findings from these tests are driving our attention towards the remarkable ability of Nature to recover from all these anthropogenic interventions. In all three studies we have faced unexpected results.

## 10.2. The Biomonitoring of the Polluted Environment in Kuwait after the Gulf War

In two species (*Arnebia decumbers* and *Plantago amplexicaulis*) we recorded a significant increase of pollen abortivity of the samples from the polluted areas compared to the control areas, while in contrast, two other species (*Neurada procumbens* and *Brassica tournifortii*) showed a decreasing abortivity effect. In the rest of the samples no changes were observed. Generally, we can state that our two-year research confirmed the surprising repairing ability of the natural environment (Malallah et al. 1997).

The same observation of the enormous ability of nature to recover from environmental catastrophes was confirmed in case of air and water pollution. Johnson et al. (1991) reported no evidence for any long-term influence of enormous clouds of smog from fired oil wells on the global weather system. The detailed measurements by Readman et al. (1992) of the contamination of the water in the Persian Gulf had showed an intensive degradation of the massive oil spill. The measured values of concentration of carcinogenic polycyclic aromatic carbohydrates were relatively low and in the sediments of some areas (Bahrain) even lower than before the military conflict. Intensive self-cleaning activity of the shore waters by the cyanobacteria after the mass death of some usual species in the shore waters (*Ocypode*, *Cleistostoma*) was well documented (Sorkhoh et al. 1992).

Paradoxically, as we observed it, after the first season of pollution of the desert with oil stains, the desert was in next season greener than ever, because the oil spiked into the soil and created an underground layer that prevented water spills from sporadic rains to leak out.

## 10.3. The Retrospective Biomonitoring of the Environmental Pollution

In the polluted area of the city Žiar nad Hronom the expected increase of production compared to the year 1894 shows the presence of genotoxic factors in the environment. The effect of pollution on the pollen grains peaked in the year 1965 (up to 56% of aberrant pollen tetrads!), when aluminum production reached its peak, in comparison with the latest samples in 1998 (15-21%) confirmed by a low level of aberrations in the following years. These figures may be the result of the current decrease of the amount of emissions due to environmental policies, the decrease in overall production as well, and the possible adaptation or tolerance of the exposed plant population (Mičieta and Murín 1998 a, Mičieta et al. 1998). The ability of such sensitive parts of plants like pollen grains to adopt a significant tolerance to this type of pollution was confirmed by other studies (Holub and Ostrolucká 1993, Chudzińska et al. 2015a, b). Ďuriš et al. (2007) as well as Križáni et al. (2007) recently reported about unexpected revitalization of the life in the worst polluted areas in Ostrava by the mining production that included the return of the species already expected to be extinct.

## 10.4. The Bioindication of Radioactive-Contaminated Areas

Although the dose of some samples of radioactive soil samples was relatively high (322 $kBq.kg^{-1}$), no significant impact on the biological level of tested wild plant species was observed (Mičieta and Murín 2007). It is important to note that the chronic effect of the radioactive pollution within 20 years and hence the very probable adaptation answer of the tested cells and plant organisms have to be taken in account. For example Zaka et al. (2002) observed within almost a half-century of the exposure to a low dose rate of ionizing radiation at the Semipalatinsk nuclear test site in Kazakhstan natural selection of the most adapted genotypes of *Stipa capillata*. This is supported by report of Oudalova et al. (2005) stating that possible adaptive response to the long-time exposition cannot be neglected too if we know that the chronic effect of radioactivity may lead to the processes of the radioresistance.

## FINAL CONCLUSION AND PERSPECTIVES

After all these findings we may refer to the older reports already questioning if the polluted environment can directly harm the human population. For example, Bates (1992) stated: "London December, 1952, smog episode. Mortality in 1952 began to increase before the sharp rise in air pollution and stayed high long after the fog cleared; there may have been underlying disease epidemic at the time." This observation was later confirmed by Anderson et al. (1994): "Studies on health effect of nitrogen dioxide have generally been equivocal, although an episode of severe nitrogen dioxide pollution in London, December, 1991 seems to have been associated with a small excess mortality." Finally, group of authors led by Sracci (1991) came to the conclusion that a critical review of relevant scientific reports cannot support the hypothesis that the polluted environment can cause a significant increase of the human life loses. Based on our experience we may only conclude that thanks to auto-regulative mechanisms, adaptation and resistance Nature showed an active self-defence against anthropogenic interventions.

Other chapters of this study are showing an alternative method of testing environmental pollution by means of plant species as bioindicators and a possibility for a recovery of affected areas by means of remediation.

# REFERENCES


Aboal, J. R., Fernández, J.A. and Carballeira, A. (2004). Oak leaves and pine needles as biomonitors of airborne trace elements pollution. *Env. and Exp. Botany*, 51, 215-225.

Abramov, V. J., Schevchenko, V. V. (1992). Genetic consequences of radioactive contamination for populations of *Arabidopsis*. *Sci. Total Environ.*, 112: 19-29.

Alfheim, I, Wikström, L. (1984). Air pollution from aluminium smelting plants I. The emission of polycyclic aromatic hydrocarbons and of mutagens from an aluminium smelting plant using the Söderberg process. *Toxicol. Environ. Chemistry*, 21: 617-28.

Al-Zalzaleh, H., Shabbir, G. (2004). Effect of bioremediated soil on growth of different plant species. *Kuwait Journal of Science and Engineering*, 31(13): 107-108.

Amadi, A., Dickson, A. A. and Maate, G. O. (1993). Remediation of oil polluted soils: 1. Effect of organic and inorganic nutrient supplements on the performance of maize (*Zea mays* L.). *Water, Air and Soil Pollution,* 66: 59-76.

Amphlett, G. E., Delow, G. (1984). Statistical analysis of the micronucleus test. *Mutat. Res.*, 128: 161-166.

Anderson, H. R., Limb, E. S., Bland, J. M., Bower, J., Ponce de Lean, A. and Strachan D.P. (1994). Health effects of an episode of nitrogen dioxide air pollution in London, *Thorax*, 49: 1053.

Aroutiounian, R. M., Ervandyan, S. G. and Nebish, A. A. (2004a). Application of male gametophyte generation of vines for the investigation of genotoxicity of environmental factors. *Proceedings of the International Conference URM. International Science and Technology Center.* Armenia, Yerevan, 46-48.

Aroutiounian, R. M., Ervandyan, S. G. and Nebish, A. A. (2004b). Estimates of total genotoxic effect of environmental factors by the analysis of male gametophyte generation of fruit trees. In: *Proceedings of the International Conference URM 2004.* International Science and Technology Center. Armenia, Yerevan, 52-54.

Baldantoni, D., De Nicola, F. and Alfani, A. (2014). Air biomonitoring of heavy metals and polycyclic aromatic hydrocarbons near a cement plant. *Atmospheric Pollution Research*, 5: 262-269.

Baležentiene, L. (2011). Analysis of species indicator values in response to spatial environmental variation of differently managed agro-habitats. *Notulae Botanicae Horti Agrobotanici Cluj-Napoca,* 39(2): 18-27.

Bezrukov, V. F., Lazarenko, L. M. (2002). Environmental impact on age-related dynamics of karyotypical instability in plants. *Mutation Research*, 520: 113-118.

Bodini, A. (2012). Building a systemic environmental monitoring and indicators for sustainability: What has the ecological network approach to offer? *Ecol. Indicators*., 15: 140–148.

Baranec, T. (1979). Karyological study of some species of genus *Pinus* L. *Biológia* (Bratislava), 34: 3-13.

Bates, D. V. (1992). Health Indices of the Adverse Effects of Air Pollution: The Question of Coherence. *Env. Res*., 59: 336-49.

Bavcon, J., Druškevič, B. and Papeš, D. (1992). *Germination of seeds and cytogenetic analysis of the spruce in differently polluted areas of Slovenia*. Phyton Austria., 33: 267-277.

enzoni, G., Viegi, L., Stefani, A., Onnis, A. (1990). Different in ro germination responses in *Pinus pinea* pollen from two localities h different levels of pollution. *Ann. Bot. Fennici*, 27: 85-90.

borty, R., Mukherjee, A. (2009). Mutagenicity and genotoxicity of l fly ash water leachate. *Ecotox. Environ. Safe.*, 72(3): 838-842.

el, J. (1997). Phytoremediation of TCE in groundwater using pulus. *U.S. EPA*, Homepage: http://clu-in.org/product/phytotce.htm.

vá, J., Mičieta, K. (2007). Peľ ako súčasť a indikátor kvality zdušia – príklad mesta Bratislavy. in: *Ovzduší 2007*, Brno, 158-162. llen as a part and indicator of air quality - an example of the city of atislava, in: *Air 2007*]

vá, J., Mičieta, K. and Ščevková, J. (2010). Monitoring of *Ambrosia* llen concentration in the atmosphere of Bratislava (Slovakia) during ars 2002-2007. *Aerobiologia,* 26(2): 83-88.

eňová, M., Karásková, P., Kallenborn, R., Klemmová Gregušková, and Čupr, P. (2016a). Pine needles for screening of Perfluorinated kylated Substances (PFASs) along ski tracks. *Environmental ience and Technology* 50(17): 9487-96.

eňová, M., Klemmová-Gregušková, E., Karásková, P., Přibylová, P., kučka, P., Baráková, D. and Čupr, P. (2016b). Pine needles d pollen grains of *Pinus mugo* Turra – A biomonitoring tool in high ountain habitats identifying environmental contamination. *nivronmental Indicators*, 132-142.

ińska, E., Diatta, J. and Wojnicka-Półtorak, A. (2014a). Adaptation rategies and referential trial of Scots and black pine populations bjected to heavy metal pollution. *Environ. Sci. Pollut. Res.*, 21: 65-2177.

ińska, E., Pawlaczyk, E. M., Celiński, K. and Diatta, J. (2014b). esponse of Scots pine (*Pinus sylvestris* L.) to stress induced by fferent types of pollutants – testing the fluctuating asymmetry. *Water nd Environment Journal,* 28(4): 533–539.

ińska, E., Wojnicka-Półtorak, A., Prus-Głowacki, W., Celiński, K. nd Diatta, J. B. (2015a). Adaptation Mechanisms of *Pinus sylvestris* . in Industrial Areas. In book: *Heavy Metal Contamination of Soils,*



Bavorová, H, Očadlíková, D. (1989). Cytogenetická analýza periférních lymfocytu. *Acta Hyg Epidemiol Microbiol,* 20:3-15. [Cytogenetic analysis of peripheral lymphocytes]

Bazhina, E. V., Kvitko, O. V. and Muratova, E. N. (2008). Meiosis at Microsporogenesis in Siberian Fir (*Abies sibirica* Ledeb.) in Natural Populations and in an Arboretum. *Eurasian. J. For. Res.*, 11(1): 41-49.

Beerling, D. J., Harfoot, M., Lomax, B. and Pyle, J. E. (2007). The stability of the stratospheric ozone layer during the end-Permian eruption of the Siberian Traps. *Philosophical Transactions of the Royal Society A: Mathematical, Physical and Engineering Sciences*, 365(1856): 1843-1866.

Bezrukov, V. F., Lazarenko, L. M. (2002). Environmental impact on age-related dynamics of karyotypical instability in plants. *Mutation Research*, 520: 113-118.

Bhowmik, S., Datta, B. K. (2012). *In Vitro* Pollen Germination of *Eichhornia crassipes* (Mart.) Solms: An Insight into its Preferred Mode of Clonal Reproduction. *Notulae Scientia Biologicae*, 4(2): 65-71.

Bonsignore, C. L., Helper, P. K. (1985). Caffeine inhibition of cytokinesis: dynamics of cell plate formation-deformation *in vivo*. *Protoplasma,* 129: 28-35.

Bothe, H., Słomka, A. (2017). Divergent biology of facultative heavy metal plants. *Journal of Plant Physiology*, 219: 45-61.

Borzan, Z., Schlarbaum, S. E. (1997). *Cytogenetic studies of forest trees and shrub species. Croatian forest,* Inc., Faculty of Forestry, University of Zagreb, Zagreb. 344 pp.

Bromberg, S. M. (1990). Identifying ecological indicators: An environmental monitoring and assessment program. *J. Air Poll. Cont.* Assoc., 40: 976-978.

Brutovská, E., Mičieta, K. and Sámelová, A. (2011). Indication of environmental mutagenesis with the use of higher plants specific locus systems. *Acta Environmentalica Universitatis Comenianae* (Bratislava), 19(1): 38-41.

Brutovská, E., Mičieta, K. and Sámelová, A. (2013). The use of plant specific locus systems in tests of genotoxicity. in: *Genetic Toxicology and Cancer Prevention*, Bratislava, SAV, 18.

Bu-Olayan, A. H., Thomas, B. V. (2002). Biomonitoring studies on the lead levels in mesquite (*Prosopis juliflora*) in the arid ecosystem of Kuwait. *Kuwait Journal of Science and Engineering,* 29(1): 65-73.

Bulgakov, N. G., Levič, A. P. and Maksimov, V. N. (2003).Regionaľnyj ekologičeskij kontroľna osnove biotičeskich i abiotičeskich dannych monitoringa. *Ekologičeskij monitoring*, Nižnij Novgorod, Izdateľstvo Nižegorodskogo univ., 93-259. (in Russian). [Regional ecological control bases biotic and abioticheskich dannych monitoring.]

Burclová, J. (1998). Decommissioning of NPP A-1 – HWGCR type, Technologies for gas cooled reactor decommissioning, fuel storage and waste disposal, in: *Proceedings of a technical committee meeting of International Atomic Energy Agency*, Vienna (Austria), 105-112.

Burton, M.A.S. (1989). *Guidelines for biological monitoring of environmental contaminants using plants*. GEMS monitoring and assessment research center. King's College London, University of London, 1-4.

Butorina, A. K., Kalaev, V. N., Najdenova, O. S., Myagkova, O. E., Vostricova, T. V., Mitroshina, O., Dirdina, O. V. and Polyakova, N. A. (2000a). Relationship between cytogenetic anomalies in forest trees subjected to radioactive contamination and industrial pollution with inherent defects in human infants in central Russia. In: Guttenberg, H., Borzan, Z., Schlarbaum, S. E. & Hartman, T. P. V. (eds.). *Cytogenetic Studies of Forest trees and Shrubs-Review, Present Status, and Outlook on the Future. Special Issue of the Forest genetics.* Arbora Publishers, Zvolen, Slovakia, 35-41.

Butorina, A. K., Kalaev, V. N., Vostrikova, T. V. and Myagokova, O. E. (2000b). Cytogenetic characteristics of seed progeny of *Quercus robur* L., *Pinus sylvestris* L. and *Betula pendula* Roth. under condition of anthropogenic contamination in Voronezh. *Tsitologiya,* 42(2): 200-201.





Butorina, A. K., Kalaev, V. N. and Karpova, S. S. mitosis and nuclear characteristics of the *Pendula* Roth.) under anthropogenous pollu 398-399.

Butorina, A. K., Kalaev, V. N., Karpova, S. damage of human somatic cells and weeping districts with different levels of anthropog *Journal of Ecology,* 33(6): 413-416.

Byrne, C. (1989). Effects of water-soluble fraction cytokinesis of the Quahong clam (*Mercen Environ. Contam. Toxico.,* 42: 81-86.

Calzoni, G.L., Antognoni, F., Pari, E., Fonti, P., Gr (2007). Active biomonitoring of heavy meta *rugosa* plants. *Environmental Pollution*, 149(2)

Capocaccia, R., Biselli, R., Ruggeri, R., Tesei, C., Rocchetti, A., Salmaso, S., Caldora, M. and Mortality in Italian veterans deployed in B Kosovo, by in: *European Journal of Public Hea*

Carneiro, M. F. H., Ribeiro, F. Q., Fernandes-Filho Barbosa, F., Rhoden, C. R., Mauad, T., S Carvalho-Oliveira, R. (2011).Pollen abortion rat passive diffusive tubes and bioaccumulation in in the characterization of air pollution. *Experimental Botany,* 72(2): 272-277.

Carvalho-Oliveira, R., Amato-Lourenço, L. F., T. C D. R. Rocha Silva,Vieira, B. D., Mauad, T., Sail Saldiva, P. H. (2016). Effectiveness of traffic-re bark and pollen abortion rates for assessing air respiratory mortality rates. *Environment Internati*

Cebulska-Wasilewska, A. (1992). *Tradescantia* bioassay on the mutagenicity of radioisot following the Chernobyl nuclear accident and *Res.,* 270: 23-29.

Springer International Publishing, Eds.: Irena Sherameti, Ajit Varma, 195-213.

Chudzińska, E., Wojnicka-Półtorak, A., Prus-Głowacki, W., Celiński, K., Diatta, J. B. and Drobek, L. (2015b). Adaptation Mechanisms of *Pinus sylvestris* L. in Industrial Areas. Heavy Metal Contamination of Soils. *Soil Biology,* Volume 44, 195-213.

Cibulka, J., Domažlická, E., Kozák, J., Kubizňáková, J., Mader, P., Machálek, E., Manikovská, B., Musil, J., Pařízek, J., Píša, J., Pohonková, H., Reisnerová, H. and Svobodová, Z. (1991). Motion of lead, cadmium and mercury in biosphere. *Academia*, Praha, 190-195.

Claxton, L. D., Woodall, G. M. Jr. (2007). A review of the mutagenicity and rodent carcinogenicity of ambient air. *Mutat. Res.*, 639: 36–94.

Clothier. B. E., Green, S. R. (1997). Roots: The big movers of water and chemical in soil. *Soil Science*, 162: 534-543.

Constantin, M. J., Owens, E. T. (1982). Introduction and perspectives of plant genetic and cytogenetic assays. *Mutat. Res.*, 99: 1-12.

Cunnigham, S. D., Berti, W. R. (1993). Remediation of contaminated soils with green plants: an overview. *In Vitro Cell. Dev. Biol.*, 29: 207-212.

Cunnigham, S. D., Berti, W. R. and Huang, J. W. (1995). Remediation of contaminated soils and sludges by green plants. In R. E. Hinchee, J. L. Means, and D. R. Burris (eds.), *Bioremediation of Inorganics*. Battelle Press, Columbus, OH, 33-54.

Dalmacija, B., Ivančev-Tumbas, I., Bikit, I., Vesković, M., Đurendić, M., Miladinov-Mikov, M., Baltić, V. V., Čonkić, L. and Bečelić, M. (2000). Environmental pollution of Novi Sad and its surroundings and health risks. *Arch. Oncol.*, 8: 113-118.

Darlington, C. D., La Cour, L. F. (1976). *The handling of chromosomes.* George Allen and Unwin Ltd., London, 208pp.

Demir, M., Makineci, E. and Gungor, B.S. (2008). Plant species recovery on a compacted skid road. *Sensors*, 8(5): 3123-3133.

De Nicola F., Baldantoni, D., Maisto G. and Alfani, A. (2015). Heavy metal and polycyclic aromatic hydrocarbon concentrations in *Quercus ilex* L. leaves fit a priori subdivision in site typologies based on human management. *Environmental Science and Pollution Research,* 24(13): 11911-11918.

De Serres, F. J. (1978). Utilization of higher plant systems as monitors of environmental mutagens. *Environ. Health Persp*., 27: 3-6.

De Serres, F. J. (1992). Higher plants as effective monitors of environmental mutagens. *Mutat. Res*, 270: 1-6.

De Simone, C., Piccola, A. and De Marco, A. (1992). Genetic effect induced by herbicide atrazine glyphosate in plant of *Vicia faba* grown in different soils. *The Science of the Total Environment*, Elsevier Science Publ. B. V., Amsterdam, 233-240.

Dienes, K., Mičieta, K. and Kubalová, S. (2009). Phytoindication and monitoring of ecogenotoxicity of dumps – the example of the recultivated dump in the village Tôň. in: *Industrial Toxicology* 2009, Bratislava, 10-14.

Dienes, K., Brutovská, E., Sámelová, A. and Mičieta, K. (2010). Porovnanie senzitivity inváznych a indikačných druhov rastlín na ekogenotoxickú deteriorizáciu. In: *Študentská vedecká konferencia. Zborník recenzovaných príspevkov*, Bratislava, Univerzita Komenského, 392-397. (in Slovak). [Comparison of the sensitivity of invasive and indicative plant species to ecogenetic deterioration. in: *Student Scientific Conference. Proceedings of Reviewed Posts*]

Dienes, K., Mičieta, K., Sámelová, A. and Brutovská, E. (2011). Detekcia vplyvu agrotechnologických postupov na ekogenotoxickú deteriorizáciu vybraných taxónov flóry Prírodnej rezervácie Mostová. *Acta Environmentalica Universitatis Comenianae* (Bratislava), 19(1): 56-59. (in Slovak). [Detection of the influence of agrotechnological practices on the ecogenotoxic deterioration of selected taxa of the Nature Reserve Mostová]

Dietz, A. C., Schnoor, J. L. (2001). Advances in Phytoremediation. *Environ. Health Perspect.*, 109: 163-168.

Ďuriš, Z., Dolný, A. and Vavříček, O. (2004). "*Paradoxy života vo vodách hornické krajiny*," Ostravská Univ., EU-Phare, 17 pp. (in Czech). ["*The paradoxes of life in the waters of the mining landscape*"]

Bavorová, H, Očadlíková, D. (1989). Cytogenetická analýza periférních lymfocytu. *Acta Hyg Epidemiol Microbiol,* 20:3-15. [Cytogenetic analysis of peripheral lymphocytes]

Bazhina, E. V., Kvitko, O. V. and Muratova, E. N. (2008). Meiosis at Microsporogenesis in Siberian Fir (*Abies sibirica* Ledeb.) in Natural Populations and in an Arboretum. *Eurasian. J. For. Res.*, 11(1): 41-49.

Beerling, D. J., Harfoot, M., Lomax, B. and Pyle, J. E. (2007). The stability of the stratospheric ozone layer during the end-Permian eruption of the Siberian Traps. *Philosophical Transactions of the Royal Society A: Mathematical, Physical and Engineering Sciences*, 365(1856): 1843-1866.

Bezrukov, V. F., Lazarenko, L. M. (2002). Environmental impact on age-related dynamics of karyotypical instability in plants. *Mutation Research*, 520: 113-118.

Bhowmik, S., Datta, B. K. (2012). *In Vitro* Pollen Germination of *Eichhornia crassipes* (Mart.) Solms: An Insight into its Preferred Mode of Clonal Reproduction. *Notulae Scientia Biologicae*, 4(2): 65-71.

Bonsignore, C. L., Helper, P. K. (1985). Caffeine inhibition of cytokinesis: dynamics of cell plate formation-deformation *in vivo. Protoplasma,* 129: 28-35.

Bothe, H., Słomka, A. (2017). Divergent biology of facultative heavy metal plants. *Journal of Plant Physiology*, 219: 45-61.

Borzan, Z., Schlarbaum, S. E. (1997). *Cytogenetic studies of forest trees and shrub species. Croatian forest,* Inc., Faculty of Forestry, University of Zagreb, Zagreb. 344 pp.

Bromberg, S. M. (1990). Identifying ecological indicators: An environmental monitoring and assessment program. *J. Air Poll. Cont.* Assoc., 40: 976-978.

Brutovská, E., Mičieta, K. and Sámelová, A. (2011). Indication of environmental mutagenesis with the use of higher plants specific locus systems. *Acta Environmentalica Universitatis Comenianae* (Bratislava), 19(1): 38-41.

Brutovská, E., Mičieta, K. and Sámelová, A. (2013). The use of plant specific locus systems in tests of genotoxicity. in: *Genetic Toxicology and Cancer Prevention*, Bratislava, SAV, 18.

Bu-Olayan, A. H., Thomas, B. V. (2002). Biomonitoring studies on the lead levels in mesquite (*Prosopis juliflora*) in the arid ecosystem of Kuwait. *Kuwait Journal of Science and Engineering,* 29(1): 65-73.

Bulgakov, N. G., Levič, A. P. and Maksimov, V. N. (2003).Regionaľnyj ekologičeskij kontroľna osnove biotičeskich i abiotičeskich dannych monitoringa. *Ekologičeskij monitoring*, Nižnij Novgorod, Izdateľstvo Nižegorodskogo univ., 93-259. (in Russian). [Regional ecological control bases biotic and abioticheskich dannych monitoring.]

Burclová, J. (1998). Decommissioning of NPP A-1 – HWGCR type, Technologies for gas cooled reactor decommissioning, fuel storage and waste disposal, in: *Proceedings of a technical committee meeting of International Atomic Energy Agency*, Vienna (Austria), 105-112.

Burton, M.A.S. (1989). *Guidelines for biological monitoring of environmental contaminants using plants*. GEMS monitoring and assessment research center. King's College London, University of London, 1-4.

Butorina, A. K., Kalaev, V. N., Najdenova, O. S., Myagkova, O. E., Vostricova, T. V., Mitroshina, O., Dirdina, O. V. and Polyakova, N. A. (2000a). Relationship between cytogenetic anomalies in forest trees subjected to radioactive contamination and industrial pollution with inherent defects in human infants in central Russia. In: Guttenberg, H., Borzan, Z., Schlarbaum, S. E. & Hartman, T. P. V. (eds.). *Cytogenetic Studies of Forest trees and Shrubs-Review, Present Status, and Outlook on the Future. Special Issue of the Forest genetics.* Arbora Publishers, Zvolen, Slovakia, 35-41.

Butorina, A. K., Kalaev, V. N., Vostrikova, T. V. and Myagokova, O. E. (2000b). Cytogenetic characteristics of seed progeny of *Quercus robur* L., *Pinus sylvestris* L. and *Betula pendula* Roth. under condition of anthropogenic contamination in Voronezh. *Tsitologiya,* 42(2): 200-201.

Butorina, A. K., Kalaev, V. N. and Karpova, S. S. (2002a). Peculiarities of mitosis and nuclear characteristics of the birch plantlets (*Betula Pendula* Roth.) under anthropogenous pollution. *Tsitologiya*, 44(4): 398-399.

Butorina, A. K., Kalaev, V. N., Karpova, S. S. (2002b). Cytogenetic damage of human somatic cells and weeping birch cells in Voronezh districts with different levels of anthropogenic pollution. *Russian Journal of Ecology,* 33(6): 413-416.

Byrne, C. (1989). Effects of water-soluble fractions of No. 2 fuel oil on the cytokinesis of the Quahong clam (*Mercenaria mercenaria*). *Bull. Environ. Contam. Toxico.,* 42: 81-86.

Calzoni, G.L., Antognoni, F., Pari, E., Fonti, P., Gnes, A. and Speranza, A. (2007). Active biomonitoring of heavy metal pollution using *Rosa rugosa* plants. *Environmental Pollution*, 149(2): 239-245.

Capocaccia, R., Biselli, R., Ruggeri, R., Tesei, C., Grande, E., Martina, L., Rocchetti, A., Salmaso, S., Caldora, M. and Francisci, S. (2016). Mortality in Italian veterans deployed in Bosnia-Herzegovina and Kosovo, by in: *European Journal of Public Health*, 26(4): 712-717.

Carneiro, M. F. H., Ribeiro, F. Q., Fernandes-Filho, F. N., Lobo, D. J. A., Barbosa, F., Rhoden, C. R., Mauad, T., Saldiva, P. H. N. and Carvalho-Oliveira, R. (2011).Pollen abortion rates, nitrogen dioxide by passive diffusive tubes and bioaccumulation in tree barks are effective in the characterization of air pollution. *Environmental and Experimental Botany,* 72(2): 272-277.

Carvalho-Oliveira, R., Amato-Lourenço, L. F., T. C. L. Moreira, T. C. L., D. R. Rocha Silva,Vieira, B. D., Mauad, T., Saiki, M. and Nascimento Saldiva, P. H. (2016). Effectiveness of traffic-related elements in tree bark and pollen abortion rates for assessing air pollution exposure on respiratory mortality rates. *Environment International*, 99: 161-169.

Cebulska-Wasilewska, A. (1992). *Tradescantia* stamen-hair mutation bioassay on the mutagenicity of radioisotope-contaminated air following the Chernobyl nuclear accident and one year later. *Mutat. Res.,* 270: 23-29.

Cela Renzoni, G., Viegi, L., Stefani, A., Onnis, A. (1990). Different in vitro germination responses in *Pinus pinea* pollen from two localities with different levels of pollution. *Ann. Bot. Fennici*, 27: 85-90.

Chakraborty, R., Mukherjee, A. (2009). Mutagenicity and genotoxicity of coal fly ash water leachate. *Ecotox. Environ. Safe.*, 72(3): 838-842.

Chappel, J. (1997). Phytoremediation of TCE in groundwater using Populus. *U.S. EPA*, Homepage: http://clu-in.org/product/phytotce.htm.

Chrenová, J., Mičieta, K. (2007). Peľ ako súčasť a indikátor kvality ovzdušia – príklad mesta Bratislavy. in: *Ovzduší 2007*, Brno, 158-162. [Pollen as a part and indicator of air quality - an example of the city of Bratislava, in: *Air 2007*]

Chrenová, J., Mičieta, K. and Ščevková, J. (2010). Monitoring of *Ambrosia* pollen concentration in the atmosphere of Bratislava (Slovakia) during years 2002-2007. *Aerobiologia,* 26(2): 83-88.

Chropeňová, M., Karásková, P., Kallenborn, R., Klemmová Gregušková, E. and Čupr, P. (2016a). Pine needles for screening of Perfluorinated Alkylated Substances (PFASs) along ski tracks. *Environmental Science and Technology* 50(17): 9487-96.

Chropeňová, M., Klemmová-Gregušková, E., Karásková, P., Přibylová, P., Kukučka, P., Baráková, D. and Čupr, P. (2016b). Pine needles and pollen grains of *Pinus mugo* Turra – A biomonitoring tool in high mountain habitats identifying environmental contamination. *Enivronmental Indicators*, 132-142.

Chudzińska, E., Diatta, J. and Wojnicka-Półtorak, A. (2014a). Adaptation strategies and referential trial of Scots and black pine populations subjected to heavy metal pollution. *Environ. Sci. Pollut. Res.*, 21: 2165-2177.

Chudzińska, E., Pawlaczyk, E. M., Celiński, K. and Diatta, J. (2014b). Response of Scots pine (*Pinus sylvestris* L.) to stress induced by different types of pollutants – testing the fluctuating asymmetry. *Water and Environment Journal,* 28(4): 533–539.

Chudzińska, E., Wojnicka-Półtorak, A., Prus-Głowacki, W., Celiński, K. and Diatta, J. B. (2015a). Adaptation Mechanisms of *Pinus sylvestris* L. in Industrial Areas. In book: *Heavy Metal Contamination of Soils,*

Springer International Publishing, Eds.: Irena Sherameti, Ajit Varma, 195-213.

Chudzińska, E., Wojnicka-Półtorak, A., Prus-Głowacki, W., Celiński, K., Diatta, J. B. and Drobek, L. (2015b). Adaptation Mechanisms of *Pinus sylvestris* L. in Industrial Areas. Heavy Metal Contamination of Soils. *Soil Biology,* Volume 44, 195-213.

Cibulka, J., Domažlická, E., Kozák, J., Kubizňáková, J., Mader, P., Machálek, E., Manikovská, B., Musil, J., Pařízek, J., Píša, J., Pohonková, H., Reisnerová, H. and Svobodová, Z. (1991). Motion of lead, cadmium and mercury in biosphere. *Academia*, Praha, 190-195.

Claxton, L. D., Woodall, G. M. Jr. (2007). A review of the mutagenicity and rodent carcinogenicity of ambient air. *Mutat. Res*., 639: 36–94.

Clothier. B. E., Green, S. R. (1997). Roots: The big movers of water and chemical in soil. *Soil Science*, 162: 534-543.

Constantin, M. J., Owens, E. T. (1982). Introduction and perspectives of plant genetic and cytogenetic assays. *Mutat. Res*., 99: 1-12.

Cunnigham, S. D., Berti, W. R. (1993). Remediation of contaminated soils with green plants: an overview. *In Vitro Cell. Dev. Biol.*, 29: 207-212.

Cunnigham, S. D., Berti, W. R. and Huang, J. W. (1995). Remediation of contaminated soils and sludges by green plants. In R. E. Hinchee, J. L. Means, and D. R. Burris (eds.), *Bioremediation of Inorganics*. Battelle Press, Columbus, OH, 33-54.

Dalmacija, B., Ivančev-Tumbas, I., Bikit, I., Vesković, M., Durendić, M., Miladinov-Mikov, M., Baltić, V. V., Čonkić, L. and Bečelić, M. (2000). Environmental pollution of Novi Sad and its surroundings and health risks. *Arch. Oncol.*, 8: 113-118.

Darlington, C. D., La Cour, L. F. (1976). *The handling of chromosomes.* George Allen and Unwin Ltd., London, 208pp.

Demir, M., Makineci, E. and Gungor, B.S. (2008). Plant species recovery on a compacted skid road. *Sensors*, 8(5): 3123-3133.

De Nicola F., Baldantoni, D., Maisto G. and Alfani, A. (2015). Heavy metal and polycyclic aromatic hydrocarbon concentrations in *Quercus ilex* L. leaves fit a priori subdivision in site typologies based on human management. *Environmental Science and Pollution Research,* 24(13): 11911-11918.

De Serres, F. J. (1978). Utilization of higher plant systems as monitors of environmental mutagens. *Environ. Health Persp.*, 27: 3-6.

De Serres, F. J. (1992). Higher plants as effective monitors of environmental mutagens. *Mutat. Res*, 270: 1-6.

De Simone, C., Piccola, A. and De Marco, A. (1992). Genetic effect induced by herbicide atrazine glyphosate in plant of *Vicia faba* grown in different soils. *The Science of the Total Environment*, Elsevier Science Publ. B. V., Amsterdam, 233-240.

Dienes, K., Mičieta, K. and Kubalová, S. (2009). Phytoindication and monitoring of ecogenotoxicity of dumps – the example of the recultivated dump in the village Tôň. in: *Industrial Toxicology* 2009, Bratislava, 10-14.

Dienes, K., Brutovská, E., Sámelová, A. and Mičieta, K. (2010). Porovnanie senzitivity inváznych a indikačných druhov rastlín na ekogenotoxickú deteriorizáciu. In: *Študentská vedecká konferencia. Zborník recenzovaných príspevkov*, Bratislava, Univerzita Komenského, 392-397. (in Slovak). [Comparison of the sensitivity of invasive and indicative plant species to ecogenetic deterioration. in: *Student Scientific Conference. Proceedings of Reviewed Posts*]

Dienes, K., Mičieta, K., Sámelová, A. and Brutovská, E. (2011). Detekcia vplyvu agrotechnologických postupov na ekogenotoxickú deteriorizáciu vybraných taxónov flóry Prírodnej rezervácie Mostová. *Acta Environmentalica Universitatis Comenianae* (Bratislava), 19(1): 56-59. (in Slovak). [Detection of the influence of agrotechnological practices on the ecogenotoxic deterioration of selected taxa of the Nature Reserve Mostová]

Dietz, A. C., Schnoor, J. L. (2001). Advances in Phytoremediation. *Environ. Health Perspect.*, 109: 163-168.

Ďuriš, Z., Dolný, A. and Vavříček, O. (2004). "*Paradoxy života vo vodách hornické krajiny*," Ostravská Univ., EU-Phare, 17 pp. (in Czech). ["*The paradoxes of life in the waters of the mining landscape*"]

Dušička, J., Mičieta, K., Brutovská, E., Sámelová, A., Ščevková, J., Zamečníková, M. and Terenová, A. (2013). Aeropalynological aspects in the detection of the quality of air in Bratislava. *Ekológia* (Bratislava), 32(1): 39-53.

Ebong, G. A., Etuk, H. S. and Johnson, A.S. (2007). Heavy metals accumulation by *Talinum triangule* grown on waste dumpsites in Uyo Metropolis, Awa Ibom State, Nigeria. *Journal of Applied Sciences*, 7(10): 1404-1409.

Ebong, G. A., Akpan, M. M. and Mkpenie, V. N. (2008). Heavy metal contents of municipal and rural dumpsite soils and rate of accumulation by *Carica papaya* and *Talinum triangulare* in Uyo, Nigeria. *E-Journal of Chemistry*, 5(2): 281-290.

Ebringer, L., Krajčovič, J. (1994). *Cell Origin and Evolution*. Publishing house VEDA, Bratislava, 321 pp.

Falk-Petersen, I. B. (1979). Toxic effects of aqueous extracts of Ekofisk crude oil, crude oil fractions, and commercial oil product on the development of sea urchin eggs. *Sarsai*, 64: 161-169.

Fargašová, A. (2009). *Ekotoxikologické biotesty*. Perfekt, Bratislava, 309 pp. (in Slovak). [*Ecotoxicological bioassays*.]

Fargašová, A. (2012). Plants as models for chromium and nickel risk assessment. *Ecotoxicology*, Volume 21, 5: 1476-1483.

Feketeová, Z., Hulejová-Sládkovičová, V., Mangová, B., Pogányová, A., Šimkovic, I. and Krumpál, M. (2016). Biological properties of extremely acidic cyanide-laced mining waste. *Ecotoxicology*, 25(1): 202-212.

Filipiak, P., Racki, G. (2010). Proliferation of abnormal palyno flora during the end-Devonian biotic crisis. *Geological Quareterly*, 54: 1-14.

Fiskesjö, G. (1985). The *Allium* test as a standard in environmental monitoring. *Hereditas*, 102: 99-102.

Fleck, A. de S., Vieira, M., Amantéa, S. L. and Rhoden, C. R. (2014). A Comparison of the Human Buccal Cell Assay and the Pollen Abortion Assay in Assessing Genotoxicity in an Urban-Rural Gradient. *International Journal of Environmental Research and Public Health* 09, 11(9): 8825-8838.

Fleck, A. de S., Carneiro, M. F. H., Barbosa Jr. F., Valladão Thiesen, F., Amantea, S. L.and Rhoden, C. R. (2016a).Monitoring an outdoor smoking area by means of $PM_{2.5}$ measurement and vegetal biomonitoring. *Environmental Science and Pollution Research*, 23(21): 21187–21194.

Fleck, A., de S., Moresco, M. B. and Rhoden, C. R. (2016b). Assessing the genotoxicity of traffic-related air pollutants by means of plant biomonitoring in cities of a Brazilian metropolitan area crossed by major highway. *Atmospheric Pollution Research*, 7(3): 488-493.

Fleischer, P., Ostrolúcka, M. and Ludvová, A. (2000). Influence of environmental factors on generative reproductive ability of *Picea abies* L. Karbs. In the Tatra National Park., *Ekológia (Bratislava),* 19(2): 117-124.

Frančová, K., Macek, T., Demnerová, K. and Macková, M. (2001). Transgenic plants – a potential tool for decontamination of environmental pollutants. *Chem. Listy*, 95: 630-637.

Gajdošová, A., Libiakova, G. and Húska, J. (2004). Improvement of selected *Amaranthus* cultivars by means of mutation techniques and biotechnological approaches. In: *Genetic improvement of under-utilized and neglected crops in low income food deficit countries through irradiation and related techniques*. IAEA-TECDOC-1426. IAEA, Vienna, Austria, 25-36.

Gajdošová, A., Libiakova, G. and Fejer, J. (2007). Improvement of selected *Amaranthus* cultivars by means of mutation techniques and biotechnological approaches. In: Sergio Ochatt and S. Mohan Jain (eds.). *Breeding of Neglected and Under-utilized Crops Spices and Herbs*, Science Publishers, USA, 151-169.

Gajdošová, A., Libiakova, G. and Fejer, J. (2007). Improvement of selected *Amaranthus* cultivars by means of mutation techniques and biotechnological approaches. In: Sergio Ochatt and S. Mohan Jain (eds.). *Breeding of Neglected and Under-utilized Crops Spices and Herbs*, Science Publishers, US, 151-169.

Garnier-Laplace, J., Geras'kin, S. A., Della-Vedova, C., Beaugelin-Seiller, K., Hinton, T. G., Real, A. and Oudalova, A. (2013). Are radiosensitivity data derived from natural field conditions consistent with data from controlled exposures? A case study of Chernobyl wildlife chronically exposed to low dose rates. *Journal of Environmental Radioactivity*, 121: 12-21.

Geras'kin, S. A., Dikarev, V. G., Zyablitskaya, Ye. Ya., Spirin, Ye. and Alexakhin, R. M. (2003). Genetic consequences of radioactive contamination by the Chernobyl fallout to agricultural crops. *J. Environ. Radioactivity*, 66: 155-169.

Geras'kin, S. A., Kim, J. K., Oudalova, A. A., Vasiliyev, D. V., Dikareva, N. S., Zimin, V. L. and Dikarev, V. G. (2005a). Bio-monitoring the genotoxicity of populations of *Scots pine* in the vicinity of a radioactive waste storage facility. *Mutat. Res.*, 583: 55-66.

Geras'kin, S. A., Vasiliyev, D. V. and Dikarev, V. G. (2005b). Bioindication-based estimation of technogenic impact on *Pinus sylvestris* L. populations in the vicinity of a radioactive waste storage facility. *Russian Journal of Ecology*, 36(4): 249-258.

Geras'kin, S., Evseeva, T. and Oudalova, A. (2011a). Plants as a Tool for the Environmental Health Assessment. *Encyclopedia of Environmental Health.* Burlington: Elsevier, 4: 571-579.

Geras'kin, S., Oudalova, A., Dikareva, N., Spiridonov, S., Hinton, T., Chemonog, E. and Garnier-Laplace, J. (2011b). Effects of radioactive contamination on *Scots pines* in the remote period after the Chernobyl accident. *Ecotoxicology,* 20(6): 1195-1208.

Geras'kin, S., Oudalova, A., Michalik, B., Dikareva, N. and Dikarev, V. (2011c). Geno-toxicity assay of sediment and water samples from the Upper Silesia post-mining areas, Poland by means of Allium-test. *Chemosphere,* 83: 1133–1146.

Geras'kin, S. A., Oudalova, A. A., Dikarev, V. G., Dikareva, N. S., Mozolin, E. M., Hinton, T., Spiridonov, S. I., Copplestone, D. and Garnier-Laplace, J. (2012). Effects of chronic exposure in populations of *Koeleria gracilis* Pers. from the Semipalatinsk nuclear test site, Kazakhstan. *Journal of Environmental Radioactivity*, 104(1): 55-63.

Geras'kin, S. A., Evseeva, T. and Oudalova, A. (2013). Effects of long-term chronic exposure to radionuclides in plant populations. *Journal of Environmental Radioactivity,* 121: 22-32.

Geras'kin, S., Vasiliyev, D. V. and Kuzmenkov, A. G. (2015). Specific Features of Scots Pine Seeds Formation in the Remote Period after the Chernobyl NPP Accident. *Radiatsionnaia biologiia, radioecologia,* 55(5): 539–547.

Geras'kin, S. A., Dikareva, N. S., Oudalova, A. A., Vasil'ev, D. V. and Volkova, P. Yu. (2016). The Consequences of Chronic Radiation Exposure of *Scots Pine* in the Remote Period after the Chernobyl Accident. *Russian Journal of Ecology*, 47(1): 26–38.

Geras'kin, S., Churyukin, R. and Volkova, P. (2017). Radiation exposure of barley seeds can modify the early stages of plants' development. *Journal of Environmental Radioactivity*, 177: 71-83.

Geras'kin, S. A., Oudalova, A., Kuzmenkov, A. and Vasil'ev, D. V. (2018). Chronic radiation exposure modifies temporal dynamics of cytogenetic but not reproductive indicators in Scots pine populations. *Environmental Pollution*, 239: 399-407.

Gichner, T., Badayev, S. A., Demchenko, S. I., Relichová, J., Sandhu, S. S., Usmanov, P. D., Usnmanova, O. and Velemínský, J. (1994). Arabidopsis assay for mutagenicity. *Mutat. Res*., 310: 249–256.

Glass, D. J. (1999). Phytoremediation: A versatile technology with many potential applications. *U.S. and International Phytoremediation*, Glass Associates Inc., Needham, 33 pp.

Gomez-Pando, L. (2014). Development of improved varieties of native grains through radiation-induced mutagenesis In: *Mutagenesis: exploring novel genes and pathways* (Eds. N.B. Tomlekova, M.I. Kozgar and M. R. Wani), Wageningen Academic Pbl., 105-123.

Gottardini, E., Cristofori, A., Cristofolini, F., Maccherini, S. and Ferretti, M. (2008). Ambient levels of nitrogen dioxide ($NO_2$) may reduce pollen viability in Austrian pine (*Pinus nigra* Arnold) trees – Correlative evidence from a field study. *Science of the Total Environment*, 402(2-3): 299-305.

Grant, W. F. (1982) Chromosome aberration assays in *Allium*. *Mutat. Res.*, 99: 273-91.

Grant, W. F. (1998). Higher plant assays for the detection of genotoxicity in air polluted environment. *Ecosyst. Health.*, 4: 210–229.

Grant, W.F. (1994). The present status higher plant bioassays for the detection of environmental mutagens. *Mutat. Res.*, 310:175-185.

Grant, W. F. (1999). Higher plants assay for the detection of chromosomal aberrations and gene mutations – a brief historical background on their use for screening environmental chemicals. *Mutat. Res.*, 310: 107-112.

Grant, W. F., Zura, K. D. (1982). Plants as sensitive *in situ* detectors atmospheric mutagens, in: Heddle, J. A. (ed.), *Mutagenicity-horizon in genetic toxicology*, Academic Press, New York, 407-434.

Grant, W. F., Lee H. G., Logan D. M. and Salamone M. F. (1992). The use of *Tradescantia* and *Vicia faba* bioassays for the *in situ* detection of mutagens in an aquatic environment. *Mutat. Res.*, 270: 53-64.

Grant, W. F., Zinovieva-Stahevitch, A. E. and Zura, K. D. (1981). Plant genetic test systems for the detection of chemical mutagens. In: H.H. Stich, R. H. C. San, (eds) *Short-Term tests for the Detection of Chemical Mutagens*, Springer, New York, 200-216.

Gregušková, E., Mičieta, K. (2013). Phytoindication of the Ecogenotoxic Effects of Vehicle Emissions Using Pollen Abortion Test with Native Flora. *Pol. J. Environ. Stud.*, 22(4): 1069-1076.

Gulis, G. (1991). Czechoslovakia: 1977 meltdown. *WISE News Communique*. Nuclear Monitor Issue: 345.

Hraška, S., Tóth, D. and Kraič, J. (2000). Monitoring of barley reserve proteins in the localities of atomic power plants during their construction and operation. *Biológia,* 55(3): 267 – 276.

Helper, P. K., Bonsignore, C. L. (1990). Caffeine inhibition of cytokinesis: ultrastructure of cell plate formation/deformation. *Protoplasma,* 157: 182-192.

Holsen, T. M., Chaberski, C. M., Khalii, N. R., Scheff, R. A. and Keil, C. B. (1991). The composition of landfill gas and its impact on local ambient quality. Report of Project OSWR-02-006. Office of Solid

Waste Research of University of Illinois, *Centre of Solid Waste Management and Research*, 32.

Holub, Z., Ostrolucká, M.G. (1993). Tolerance of pollen from trees to chronic effect of Pb and acidity. *Biológia (Bratislava),* 48: 331-335.

Holub, Z., Zeleňáková, E. (1985). Problems of stabilizing the heaps of nickeliferous ore tailings by a vegetative cover. Experiments with *Dactylis glomerata*. *Biológia (Bratislava),* 40: 497-504.

Holub, Z., Banásová, V. and Zeleňáková E. (1987). The dynamic structure and heavy metal contamination in vegetation under long term influence of Pb and Cu emissions. *Ekológia (Bratislava),* 6: 101-111.

Hölgye, Z., Filgas, R. (1993). Monitoring of the release of transuranium elements in airborne effluents from V-1 nuclear power plant at Jaslovské Bohunice in Czechoslovakia. *J. Environ. Radioact.*, 20: 191-199.

Hool, K. D., Cairns, J. (1995). Landscape indicators in ecotoxicology. In: Hoffman, D. J., Rattner, A. B., Burton, G. A. and Cairns, J. (eds.). *Handbook of ecotoxikology*. Lewis Publishers, London-Tokyo, 185-197.

Houghton, R. A. (1995). Global effects of deforestration. In: Hoffman, D. J., Rattner, A. B., Burton, G.A. and Cairns, J. (eds.). *Handbook of ecotoxikology.* Lewis Publishers, London-Tokyo, 492-508.

Houston, D. B., Dochinger, L.S. (1977). Effect of ambient air pollution on cone, seed and pollen characteristic in eastern white and red pines. *Environ. Pollut.*, 12: 1-5.

Hricová, A., Libiaková, G. and Gajdošová, A. (2010). Breeding of amaranth through radiation mutagenesis. Nové poznatky z genetiky a šľachtenia poľnohospodárskych rastlín. In: *Zborník zo 17. vedeckej konferencie, Piešťany: VÚRV*, 116–117. [New knowledge of genetics and breeding of agricultural plants. In: *Proceedings of the 17th Scientific Conference*, Piešťany: VÚRV]

Hricová, A., Kečkešová, M., Gálová, Z., Libiaková G., Gajdošová, A. (2011). Skúmanie zmien profilu bielkovín v semenách láskavca podrobených radiačnej mutagenéze. *Chemické listy*, 105: 542-545. (in

Slovak) [Examining changes in the protein profile in the seeds of the subject subject to radiation mutagenesis. *Chemical Sheets*]

IAEA (1988). *The Radiological Accident in Goiania. Vienna*, 157 pp.

Ichikawa S. (1992). *Tradescantia* stamen-hair system as an excellent botanical tester of mutagenicity: its responses to ionizing radiations and chemical mutagens, and some synergistic effects found. *Mutat. Res.*, 270: 3-22.

Isaenkov, S. V., Sorochinskii, B. V., Sokolov, N. V., Grodzinskii, D. M. (1999). Radiation-induced changes in the genome of pine seeds, forming under conditions of chronic irradiation in the Chernobyl exclusion zone. *Ukr. Biokhim. Zh.*, 71(4): 103-6.

ISCN (1995). *An International System for Human Cytogenetic Nomenclature.* Karger, Basel. 66-76.

Jadczyk, P. (2003). Methods of examination of genotoxicity of atmospheric pollutants. *Environment Protection Engineering*, 29(3-4): 97-111.

Jančovičová, S., Zaliberová, M. (2011). Macromycetes near the city incinerator in Bratislava. *Catathelasma*, 13: 16.

Javorka, K., Javorka, Ľ., Halaša, M. and Baranec, M. (2005). Hodnotenie spontánnej potratovosti v spádovej oblasti okresu Ružomberok v období rokov 1998-2003. *Disputationes Scientificae Universitatis Catholicae in Ružomberok*, 5(1): 52-57. [Assessment of spontaneous abortion in the catchment area of the Ružomberok district in 1998-2003.]

Jedrychowski, W. A., Perera, F. and Maugeri, U. (1998). Molecular epidemiology in preventive medicine, *Int. Center for Studies and research in Biomedicine,* Luxembourg, 273 – 280.

Johnson, D. W., Kilsby, C.G., McKenna, D.S., Saunders, R.W., Jenkins, G. J., Smith, F. B. and Foot, J. S. (1991). Airborne observations of the physical and chemical characteristics of the Kuwait oil smoke plume. *Nature,* 353: 617-621.

Jordahl, J. L., Foster, l., Schnoor, J. L. and Alvarez, P. J. J. (1997). Effect of hybrid poplar trees on microbial populations important to hazardous waste bioremediation. *Environ. Toxicol. And Chemistry*, 16: 1318-1321.

Kalaev, V. N., Karpova, S. S. (2003). The Influence of Air Pollution on Cytogenetic Characteristics of Birch Seed Progeny. *Forest Genetics,* 10(1): 11-18.

Kanaya, N., Gill, B.S., Grover, I. S., Murín, A., Osiecka, R., Sandhu, S.S. and Andersson H. C. (1994). *Vicia faba* chromosomal aberration assay. *Mutat. Res.*, 310: 231-247.

Kandziora-Ciupa, M., Ciepał, R., Nadgórska-Socha, A. and Barczyk, G. (2016). Accumulation of heavy metals and antioxidant responses in *Pinus sylvestris* L. needles in polluted and non-polluted sites. *Ecotoxicology,* 33-45.

Kappler, R., Kristen, U. (1987). Photometric quantification of in vitro pollen tube growth: a new method suited to determine the cytotoxicity of various environmental substances. *Environ. Exp. Botany*, 27: 305-309.

Karunakara, N., Somashekarappa, H. M., Narayana, Y., Avadhani, D. N., Mahesh, H. M. and Siddappa, K. (2003). $^{226}$Ra, $^{40}$K and $^{7}$Be activity concentrations in plants in the environment of Kaiga, India. *J. Environ. Radioactivity*, 65: 255-266.

Käffer, M. I., Lemos, A. T., Apel, M. A., Rocha, J. V., Martinus, S. M. A. and Vargas, V. M. F. (2012). Use of bioindicators to evaluate air quality and genotoxic compounds in an urban environment in Southern Brazil. *Environ. Pollut.*, 163: 24–31.

Kearns, C. A., Inouye, D. W. (1993). *Techniques for Pollination Biologists.* Univ Press, Colorado, 583pp.

Klubicová, K., Danchenko, M., Škultéty, Ľ., Miernyk, J.A., Rashydov, N.M., Berezhna, V. V., Preťová, A. and Hajdúch, M. (2010). Proteomics analysis of flax grown in Chernobyl area suggests limited effect of contaminated environment on seed proteome. *Environmental science & technology*, 44(18): 6940-6946.

Klubicová, K., Berčák, M., Danchenko, M., Škultéty, Ľ., Rashydov, N.M. and Berezhna, V.V. (2011). Agricultural recovery of a formerly radioactive area: I. Establishment of high-resolution quantitative protein map of mature flax seeds harvested from the remediated Chernobyl area. *Phytochemistry,* 72: 1308–1315.

Klubicová, K., Vesel, M., Rashydov, N.M. and Hajdúch, M. (2012). Seeds in Chernobyl: the database on proteome response on radioactive environment. *Front Plant Science*, 3: 231-235.

Knasmüller, S., Kim, T-W. and Ma, T-H. (1992). Synergistic effect between tannic acid and X-rays detected by the *Tradescantia*-micronucleus assay. *Mutat. Res.,* 270: 31-37.

Koster, A. S. J., Van den Biggelaar, J. A. M. (1980). Abnormal development of *Dentalium* due to the Amoco Cadiz oil spill. *Mar. Pollut. Bull.*, 11: 166-169.

Kovacs, M. (1992). Biological indicators in environmental protection. *Ellis Harwood series in environmental management, science and technology*, 7-11.

Kovalchuk, I., Kovalchuk, O., Arkhipov, A. and Hohn, B. (1998). Transgenic plants are sensitive bioindicators of nuclear pollution caused by the Chernobyl accident. *Nature Biotechnology*, 16: 1054-1057.

Krahulcová, A., Jarolímová, V. (1993). Ecology of two cytotypes of *Butomus umbellatus*. I. Karyology and breeding behaviour. *Folia Geobot. Phytotax.* (Praha), 28: 385-411.

Kristen, U. (1997). Use of higher plants as screens for toxicity assessment. *Toxicol. In Vitro*, 11: 181-191.

Kristen, U., Kappler, R. (1995). The pollen tube growth test. In: O'Hare, S., Atterwill, C. K. (eds.). *In vitro toxicity testing protocols. Methods in molecular biology.* Humana Press, Totowa, New York, 189-198.

Križáni, I., Andráš, P. and Ladomerský, J. (2007). Banícke záťaže Štiavnických vrchov. [Mining loads of the Štiavnických mountains.] Vydavateľstvo Technickej univerzity vo Zvolene, 100pp.

Kuchma, O., Finkeldey, R. (2011). Evidence for selection in response to radiation exposure: *Pinus sylvestris* in the Chernobyl exclusion zone. *Environmental Pollution*, 159(6): 1606-1612.

Kuchma, O., Vornam, B. and Finkeldey, R. (2011). Mutation rates in Scots pine (*Pinus sylvestris* L.) from the Chernobyl exclusion zone evaluated with amplified fragment-length polymorphisms (AFLPs) and microsatellite markers. *Mutation Research – Genetic Toxicology and Environmental Mutagenesis*, 725(1-2): 29-35.

Kumar, S., Bansal, H. C., Singh, D. and Natarajan, A. T. (1966). Consistency of karyotypes and classification and identification of chromosomes in the genus *Pinus*. *Indian J. Genet. Plant Breed.*, 26: 311-316.

Kunová, K. (1999). Bioindication of phytotoxicity and genotoxicity through higher plants in the ore mine area. *Ekol. Bratislava*, 3: 271-276.

Kunová-Staples, K., Mičieta, K. (2000). Selected aspects for evaluation of phytotoxicity and genotoxicity of heavy metals. In: Anke, M., (ed.), *Mengen- und spurenelemente*, Schubert-Verlag, Leipzig, 617-627.

Kunová-Staples, K., Mičieta, K. (2002). Heavy metals in the environment and phytoindication of their genotoxic effects at the ore mine area. *Acta Fac. Rerum Nat. Univ. Comenianae, Bot.*, 41: 15-21.

Lakes, T., Kim, H. O. (2012). The urban environmental indicator "Biotope Area Ratio" - An enhanced approach to assess and manage the urban ecosystem services using high resolution remote-sensing. *Ecol. Indic.,* 13: 93–103.

Letko, V., Sergejev, V. I. and Šimko, T. G. (1992). "*Research of geological surrounding pollution from sludge beds of Pezinok ore mine*" (geological work). Department of Engineering geology, Faculty of Natural Sciences, Comenius University, Bratislava, 45-48.

Letkovičová, M., Korec, J., Príkazský, V. and Smejkal, Z. (1997). Twenty years' development of the incidence of cancer illness in the vicinity of nuclear power plant at Jaslovské Bohunice. *Journal of Radioanalytical and Nuclear Chemistry*, 218(1): 111-115.

Liman, R., Ciĝerci, H. I., Akyil, D., Eren, Y. and Konuk, M. (2011). Determination of genotoxicity of fenaminosulfate by *Allium* and Comet tests. *Pestic. Biochem. Phys.*, 99(1): 61-64.

Loppi, S., Riccobono, F., Zhang, Z. H., Savić, S., Ivanov, D. and Pirintsos, S.A. (2003). Lichens as biomonitors of uranium in the Balkan area. *Environmental Pollution*, 125(2): 277-280.

Lourenço, J., Mendo, S. and Pereira, R. (2016). Radioactively contaminated areas: Bioindicator species and biomarkers of effect in an early warning scheme for a preliminary risk assessment. *Journal of Hazardous Materials*, 111-121.

Macnair, M. R. (1987). Heavy metal tolerance in plants: A model evolutionary system. *Trends in Ecol and Evol*; 3: 354-58.

Ma, T.-H. (1982). Tradescantia cytogenetic tests (root-tip mitosis, pollen mitosis, pollen mother-cell meiosis). *Mutat. Res.,* 99: 293-302.

Ma, T-H. (1990). *In situ* monitoring of environmental clastogenes using *Tradescantia* – micronucleus bioassay. In: Sandhu S. S., Lower W. R., De Serres F. J., Suk W. A., Tice R.R. (eds) *In situ evaluation of biological hazards of environmental pollutants*. Plenum Press, New York, 183-189.

Macnair M. R. (1987). Heavy metal tolerance in plants: A model evolutionary system. *Trends in Ecology and Evolution*, 3: 354-358.

Makineci, E., Gungor, B. S. and Demir, M. (2008). Survived herbaceous plant species compacted skid road in a fir (*Abies bornmulleriana* Mattf.) forest. *Transportation Research*, 13(3): 187-192.

Malallah, G., Afzal, M., Murín, G., Murín, A. and Abraham, D. (1997). Genotoxicity of oil pollution on some species of Kuwaiti flora. *Biológia (Bratislava),* 52(1): 61-70.

Maňkovská, B., Puera R. and Huttunen S. (1988). Deposition of air-borne pollutants on the surface of spruce needles around the aluminium smelting. *Ekológia (Bratislava),* 7: 291-298.

Marques, A. P. G. C., Rangel, A. O. S. S. and Castro, P. M. L. (2009). Remediation of Heavy Metal Contaminated Soils: Phytoremediation as a Potentially Promising Clean-Up Technology. In: *Critical Reviews in Environmental Science and Technology*, 39: 622–654.

Mičieta, K. (1987). Increased aberration capacity of chromosomes in *Vicia faba* L. under the influence of a combination of some herbicides after treatment in field cultures. *Agriculture (Nitra),* 33: 1094-1100.

Mičieta, K. (1988). Increased frequency of sister chromatid exchanges for *Vicia faba* L. due to the combination of herbicides Tobogard, Dual, Ladob. *Agriculture (Nitra),* 34(5): 438-442. (in Slovak)

Mičieta K. (1989). Cultivated plants as indicators of phyto-toxicity and mutagenity of contaminated environment. *Agriculture (Nitra),* 35: 122-131.

Mičieta K. (1990). Bioindication of mutagenous effects of polluted environment by means of higher plants. *Živ. Prostr.,* 4: 267-270.

Mičieta, K. (1992a). Aspects of adaptation of higher plant bioindicators for detection genotoxicity of the polluted environment. *Průmyslová toxikologie a ekotoxikologie*, 19: 32-33. (in Slovak) [*Industrial Toxicology and Ecotoxicology*]

Mičieta K. (1992b). Bioindication of genotoxic effects of emissions of local flora species in Žiarská kotlina. In: *Ekologická únosnosť regiónu Žiarskej kotliny, Ekotrust, Banská Štiavnica*, 183-192. (in Slovak) [*Ecological bearing capacity of Žiarska kotlina region, Ekotrust, Banská Štiavnica*]

Mičieta, K. (1993). Local flora in bioindication and evaluation of genotoxic risk in air pollution areas of industrial complexes. in: *Průmyslová toxikologiea ekotoxikologie*, Pardubice, 35-36. (in Slovak) [*Industrial toxicology and ecotoxicology*]

Mičieta, K. (1998). Pollen in retro-biomonitoring of the contaminated environment. In: Benčaťová, B., Hrivnák, R. (eds.), *Rastliny a človek*, TU Zvolen, 73-76. (in Slovak) [*Plants and man*]

Mičieta, K. (2009). Peľ ako súčasť a indikátor kvality životného prostredia. In: *Progres v biológii*, UKF, Nitra, 17-22. (in Slovak) [Pollen as an indicator of the quality of the environment. *Progress in biology*]

Mičieta, K., Dušička, J. (2000). Importance of arborets in genotoxic indication and evaluation deterioration of the tree genofond. *Arboréta: premenlivosť a introdukcia drevín,* Zvolen: Lesnícky výskumný ústav, 87-91. (in Slovak) [*Arboréta: variability and introduction of woody plants*, Zvolen: Forestry Research Institute]

Mičieta, K., Gregušková, E. (2013). Phytoindication of the ecogenotoxic effects of vehicle emissions using pollen abortion test with native flora. *Pol. J. Environ. Stud.*, 22(4): 1069-1076.

Mičieta, K., Kunová, K. (2000a). Phytoindication of genotoxic deterioration of polluted environment. *Biológia (Bratislava),* 55: 75-79.

Mičieta K., Kunová K. (2000b). *In situ* bioindication and monitoring of ecotoxic effects using species of native flora in the vicinity of a nickel plant. In: *Industrial Toxicology*, 16-20.

Mičieta, K., Murín, G. (1993a). Higher plants in bioindication and monitoring of genotoxicity of the polluted environment. in: Marko, J. (ed.), *Monitoring of environment*, MŽP SR, Bratislava, 81-92.

Mičieta, K., Murín, G. (1993b). Bioindication of genotoxicity by means of species of regional flora in monitoring of polluted environment. in: Eliáš, P. (ed.), *Monitoring bioty*, Bratislava, SAV, 41-56.

Mičieta, K., Murín, G. (1994a). Bioindication of "in vivo" and "in situ" phytotoxicity and genotoxicity on contaminated sites in the vicinity of EBO Jaslovské Bohunice in its capacity use in biomonitoring the environment of nuclear power plants. Trnava: *Výskumný ústav jadrových elektrární*, 63. (in Slovak)

Mičieta, K., Murín, G. (1994b). Reflection of genotoxic environmental impact to the tetrade analysis *Calluna vulgaris* in the last 100 years in selected regions of Slovakia. In: Eliáš, P. (ed.), *Populačná biológiarastlín III.,* SEKOS, 68-70. (in Slovak)

Mičieta, K., Murín, G. (1995). Woods in bioindication and genotoxicity monitoring in a polluted environment. *Průmyslová toxikologie a ekotoxikologie*, 22: 42-44. (in Slovak)

Mičieta, K., Murín, G. (1996a). Microspore analysis for genotoxicity of a polluted environment. *Environ. Exp. Bot.*, 36(1): 21-27.

Mičieta, K., Murín, G. (1996b). Local flora and field cultures in bioindication and monitoring genotoxicity of the polluted environment. In: Kontrišová, O., Kočík, K., Bublinec, E. (eds.), *Monitorovanie a hodnotenie stavu životného prostredia*, FEETU Zvolen a ÚEL SAV Zvolen, 199-203. (in Slovak)

Mičieta, K., Murín, G. (1996c). The response of some bioindicators, species of Slovakian flora, from genotoxic impact of polluted environment. *Population Biology IV.*, 85-89.

Mičieta, K., Murín, G. (1996d). New utility of herbal samples for retrospective biomonitoring of polluted environment. *Život. prostr.*, Bratislava, 5: 262-263.

Mičieta, K., Murín, G. (1997a). Wild plant species in practical use for bioindication of polluted environment. *Ekológia (Bratislava),* 16(2): 193-202.

Mičieta, K., Murín, G. (1997b). Genus Pinus in Bioindication of Polluted Environment at selected locations in Slovakia. In: *III. Dendrologické dni (Proc.),* Nitra, 74-78. (in Slovak)

Mičieta, K., Murín, G. (1997c). Criteria for selecting Slovak flora for bioindication and genotoxicity monitoring. In: *III. Dendrologické dni* (Proc.), Nitra, 69-73. (in Slovak)

Mičieta, K., Murín, G. (1997d). The use of *Pinus sylvestris* L. and *Pinus nigra* Arnold as a bioindicator species for environmental pollution. Cytogenetic studies of forest trees and shrub species (Eds. Borzan, Ž., Schlarbaum, S.E.), *Proc. Of the First IUFRO Cytogenetics Working Party*, Univ. of Zagreb, 253-263.

Mičieta, K., Murín, G. (1998a). Different levels of plant response to environmental stresses. in: Terek, O. I. et al. (eds.), *Proceedings of International conference "Plant ontogenesis in natural and transformed environments,"* Spolom Publ., Lviv, Ukraine, 270-271.

Mičieta, K., Murín, G. (1998b). Three species of genus *Pinus* suitable as bioindicators of polluted environment. *Water Air Soil Poll.*, 104: 413-422.

Mičieta, K., Murín, G. (1998c). Dreviny v bioindikácii a monitoringu ekotoxicity mestského intravilánu. in: Benčať, T., Gregor, M. (eds.), *Porasty a dendrologické objekty Slovenského stredohoria a aktuálne problémy dendrológie*, TU Zvolen, 173-177. [Wood in bioindication and ecotoxicity monitoring of urban intravilan, in: *Porasty and dendrological objects of the Slovak Stredohoria and current problems of dendrology*]

Mičieta, K., Murín, G. (1998d). Effect of air pollution to re-introduced arborescent species. in: De Kok, L. J., Stulen, I. (eds.), *Responses of plant metabolism to air pollution and global change*, Backhuys Publ., 379-381.

Mičieta K., Murín G. (1999). Retrospective Biomonitoring: A hundred years of environmental pollution at selected areas in Slovakia. *J. Environ. Pathol. Toxicol. Oncol.*, 18(4): 335-338.

Mičieta, K., Murín, G. (2000). Seasonal and individual variability response of woods in the bioindication of genotoxicity. In: Vreštiak, (ed.), Zborník referátov zo IV. *Dendrologické dni, Adaptabilita drevín na aridizáciu prostredia, Piešťany,* 18-22. (in Slovak)

Mičieta, K., Murín, G. (2002a). Physiological indication of genotoxic deterioration in the environment. In: *25. pracovní dny České a slovenské společnosti pro mutagenezu zevním prostředím Československé biologické společnosti,* IDVPZ Brno, Brno, Czech Republic, 53-54. (in Slovak) [*25th Business Days of the Czech and Slovak Society for Mutagenesis Outside the Environment of the Czechoslovak Biological Society*]

Mičieta, K., Murín, G. (2002b). Simultánny test fytotoxicity a mutagenity. *Ecotoxicity Bioassays 1*, Kočí V. and Halousková O. (Eds.), Ecotox, Czech Republic, 147-155. (in Slovak) [*Ecotoxicity Bioassays 1*]

Mičieta, K., Murín, G. (2003a). Aspects of hypoxia in the phytoindication of ecotoxicity and mutagenicity. In: Kočí, V., Halousková, O. (eds.), *Ecotoxicity Bioassays II*, Ecotox, Praha, Czech Republic, 105-107. (in Slovak)

Mičieta, K., Murín, G. (2003b). Individual, populational and special specificity of ecogenetic deterioration. In: *26. pracovní dny České a Slovenské společnosti pro mutagenezu zevním prostředím při Československé biologické společnosti,* Národní centrum ošetřovatelství a nelékařských zdravotníckych oborů, Brno, 37-38. (in Slovak) [*26th Business Days of the Czech and Slovak Society for Mutagenesis Outside the Czechoslovak Biological Society,* National Center of Nursing and Non-medical Healthcare]

Mičieta, K., Murín, G. (2003c). Cytogenetická analýza vo fytoindikácii ekogenotoxicity v podmienkach *in situ*. In: Kočí, V., Maršálek, B., Halousková, O. (eds.), *Ecotoxicity Bioassays III*, Brno, Czech

Republic, 150-153. (in Slovak) [*Cytogenetic analysis in phytoindication of ecogenotoxicity under in situ conditions*.]

Mičieta, K., Murín, G. (2003d). Phytoindication of ecogenotoxicity in situ conditions, its current and retrospective monitoring at selected localities. *Acta Fac. Rer. Nat. Univ. Ostraviensis,* Biologica – Ecologica, 10: 147-152. (in Slovak)

Mičieta, K., Murín, G. (2004). Comparison of the sensitivity of endangered and invasive species to ecogenotoxic deterioration and the use of this aspect in phytoremediation strategies. *Ecotoxicity Bioassays IV*, Czech Republic, Kočí, V., Maršálek B. Tlustá (eds.), 161-167. (in Slovak)

Mičieta, K., Murín, G. (2007). Wild plant species in bio-indication of radioactive-contaminated sites around Jaslovské Bohunice nuclear power plant in the Slovak Republic. *J. Environ. Radioact*., 93(1): 26-37.

Mičieta, K., Murín, G. (2008). Methodology as a modifier of a genotoxic effect by means of a simultaneous phytotoxicity and mutagenicity test. In: *Aktuální problematika genetické toxikologie*, 31. pracovní dny České a slovenské společnosti pro mutagenezu zevním prostředím při Československé biologické společnosti, Národní centrum ošetřovatelství a nelékařských zdravotníckych oborů, Brno, 49-51. (in Slovak) [In: *Current Issues of Genetic Toxicology*, 31st Business Days of the Czech and Slovak Society for Mutagenesis Outside the Czechoslovak Biological Society, National Center of Nursing and Non-Medical Healthcare]

Mičieta, K., Murín, G. (2014). Adaptation to the environmental stress as a result of a long-term mutagenesis in plants. In: Tomlekova NB, Kozgar MI, Wani MR (eds) *Mutagenesis: exploring novel genes and pathways* (eds), Wageningen Academic Pbl., 379-404.

Mičieta, K., Polák, R. (2000). Phytoremediations – effective bio-technologies of the xenobiotic's elimination in the environment. *Proc. Tech. of Environ. Protection*, 7-12.

Mičieta, K., Polák, R. (2001). Phytoremediations systems in the localization and elimination nitrate compounds from the groundwater. *Proc. Tech. of Environ. Protection*, 193-198.

Mičieta, K., Ostrovský, I. and Murín, G. (1998). Native flora in the bioindication of the genotoxicity of PAH. in: Weiss, P., Schröder, P., Rether, B., Keith, G., Collins, C., Bach, Th. (eds.), *Organic xenobiotics and plants: impact, metabolism and toxicology*, Federal Environment Agency – Austria, Proc. of the IMTOX'97, Vienna, 97-104.

Mičieta, K., Murín, G. and Kunová, K. (1999a). Impact of environmental stress on plants and parallel assessment. In: Alberty, R. (ed.), *Biologic Days '99,* Univerzita Mateja Bela in Banská Bystrica, 93-95. (in Slovak).

Mičieta, K., Kunová, K. and Murín, G. (1999b). Indication methods for evaluation of acute and persistent genotoxic exposure in the condition "*in situ*." In: Romančík, V. (ed.), *Industrial Toxicology* 1999, Bratislava, 124-128.

Mičieta, K., Kunová, K., Murín, G. and Stano J. (1999). Phytoindication and monitoring of genotoxicity in the vicinity of the Bratislava incinerator. In: TOP'99 (Zborník z Medzinárodnej konferencie Technika ochrany prostredia), STU Bratislava, 69-74. (in Slovak)

Mičieta, K., Kunová, K. and Mišík, M. (2000). Comparison of sensitivity of clonal and heterozygous populations to polluted environment. In: Eliáš, P. (ed.), *Populačná biológia rastlín 6*, SEKOS, Banská Štiavnica, 69-75. (in Slovak)

Mičieta, K., Murín, G., Kunová, K., Halašová, E. and Bukovská, E. (2002a). Monitoring of genotoxic risk in industrial areas with an example of Ružomberok papermill and factory. *J. Epidemiol.*, 56(2): 15-121.

Mičieta, K., Kunová-Staples, K. and Murín, G. (2002b). Adaptation on environmental stress and its aspects in strategy of phytoremediation. *Acta Fac. Rerum Nat. Univ. Comenianae*, Bot. 41, Bratislava, 3-8.

Mičieta, K., Hutár, V. and Murín G. (2002c). Phytoindication of soil eco- and genotoxicity. *Phytopedon* (Bratislava), 1: 143-148. (in Slovak)

Mičieta K., Murín G., Kunová K., Halašová E. and Bukovská E. (2002d). Monitoring of genotoxic risk in industrial areas with an example of Ružomberok papermill and factory. *J. Epidemiol.*, 56(2): 15-121.

Mičieta, K., Murín, G. and Mišík, M. (2003a). Microsporogenesis and pollen of higher plants in the indication and monitoring of ecogentoxicity under *in situ* conditions. In: Kočí, V., Maršálek, B., Halousková, O. (eds.), *Ecotoxicity Bioassays III,* Brno, Czech Republic, 158-167. (in Slovak)

Mičieta, K., Murín, G., Mišík, M. and Vršková, K. (2003b). Simultaneous phytotoxicity and mutagenicity test of some organic solvents used in ecotoxicity bioassays. In: Kočí, V., Halousková, O. (eds.), *Ecotoxicity Bioassays II*, Ecotox, Praha, Czech Republic, 80-83. (in Slovak)

Mičieta, K., Solenská, M., Mišík, M. and Murín, G. (2004a). Validation of plant tests in the indication and monitoring of ecogenotoxicity under *in situ* conditions. In: Kočí, V., Maršálek, B., Tlustá, P. (eds.), *Ecotoxicity Bioassays IV*, Chrudim, Czech Republic, 114-123. (in Slovak)

Mičieta K., Mišík M., Solenská M. and Murín G. (2004b).Physioindication and monitoring of ecogenotoxicity of current xenobiotic exposure in industrial locations in Slovakia. In: *Monitorovanie a hodnotenie stavu životného prostredia* 5. časť A: Výsledky výskumu. Technická univerzita, Zvolen, 143-149. (in Slovak)

Mičieta, K., Solenská, M., Mišík, M. and Murín, G. (2005). Validation of phytoindication of ecogenotoxicity. Aktuální problematika genetické toxikologie, Rössner, (ed.), 38. *Výroční zasedání cytogenetické sekce,* NCO NZO Brno, Czech Republic, 43-46. (in Slovak)

Mičieta, K., Murín, G. and Chrenová, J. (2006a).Bioindication of the local flora of radioactive contaminated areas in the Jaslovské Bohunice NPP after the accident in 1977. In: *XIX. Konferencia "Radionuklidy a ionizující záření ve vodním hospodářství,"* České Budějovice, Česká republika, 35-44. (in Slovak)

Mičieta, K., Murín, G. and Mišík, M. (2006b).Phytoindication of the Mutagenicity of Radioactive Contaminated Locations and its Possibilities for Use in Biomonitoring the Environment of Nuclear Power Plants. In: *29. pracovní dny České a slovenské společnosti pro mutagenezu zevním prostředím při Československé biologické společnosti,* Národní centrum ošetřovatelství a nelékařských zdravotníckych oborů, Brno, 67-70. (in Slovak)

Mičieta, K., Pisarčíková, H., Šípková, Ž. and Murín, G. (2006c). Microsporogenesis of wood species in indication and monitoring of ecogenotoxicity. in: Kočárek, Plášek, V., Malachová, K. (eds.), *Environmental changes and biological assessment III.* Scripta Facultatis Rerum Naturalium Univesitatis Ostraviensis Nr. 163, Ostrava, 331-338.

Mičieta, K., Mišík, M. and Murín, G. (2007). Agrocenoses in the indication and monitoring of ecogenotoxicity – possibilities and limits. In: *30. prac. dny České a slovenské společnosti pro mutagenezi zevním prostředím při Československé biologické spol.*, IDVPZ, Brno, 61-64. (in Slovak) [*30. prac. Days of the Czech and Slovak Society for External Mutagenesis at Czechoslovak Biological Society*]

Mičieta, K., Dušička, J. and Murín, G. (2008a). Local dendroflora in the detection of ecogenotoxicity monitoring near the Slovnaft Bratislava industrial complex in the last twenty years. In: *Industrial Toxicology/ Priemyselná toxikológia*, Bratislava, 125-130. (in Slovak)

Mičieta, K., Murín, G. and Dušička, J. (2008b). Local flora in the detection and monitoring of ecogenotoxicity in the vicinity of motorways and highways. in: *Genetic Toxicology and Cancer Prevention*, Brno, 15-18.

Mičieta, K., Murín, G., Brutovská, E., Dušička, J., Dienes, K. and Sámelová, A. (2009a). Detection of the ecogenotoxic impact of the polluted environment on the rare and endangered plant species. In: *Genetic Toxicology and Cancer Prevention*, Bratislava, 29-32.

Mičieta, K., Dušička, J., Mišík, M. and Murín, G. (2009b). Phytoindication and Monitoring of Ecogenotoxicity *in situ* in Traffic Emissions in Bratislava. In: *Ovzduší 2009,* 120-124. (in Slovak)

Mičieta, K., Érsek, A. and Dušička, J. (2010a). *Arabidopsis thaliana* – test for the indication and monitoring of ecogenotoxicity under *in situ* conditions. In: *33. pracovní dny České a slovenské společnosti pro mutagenezi zevním prostředím při Československé biologické společnosti*, Národní centrum ošetřovatelství a nelékařských zdravotníckych oborů, Brno, 29-31. (in Slovak) [*33 working days of the Czech and Slovak society for mutagenesis outside the Czechoslovak Biological Society*]

Mičieta, K., Brutovská, E. and Sámelová, A. (2010b). Specific locus systems of higher plants as indicators for genotoxic deterioration. *Acta Bot. Univ. Comen.*, 45: 33-40.

Mičieta, K., Bikar, T. and Murín, G. (2010c). Simultaneous test of phytotoxicity and mutagenesis of contaminated sites after NATO bombing in Serbia. In: *33. pracovní dny České a slovenské společnosti pro mutagenezu zevním prostředím při Československé biologické společnosti*, Národní centrum ošetřovatelství a nelékařských zdravotníckych oborů, Brno, 26-27. (in Slovak) [*33 working days of the Czech and Slovak society for mutagenesis outside the Czechoslovak Biological Society*]

Mičieta, K., Dušička, J., Brutovská, E., Sámelová, A. and Dienes, K. (2011a). Photoindication and monitoring of ecogenotoxicity in the vicinity of motorways and transport communications. *Acta Environmentalica Universitatis Comenianae* (Bratislava), 19(1): 215-218. (in Slovak)

Mičieta, K., Dušička, J., Brutovská, E., Sámelová, A. and Dienes, K. (2011b). Detection of the influence of motorways on the genomic instability of the local flora. In: *Genetic Toxicology and Cancer Prevention*, Brno, 30-34.

Mičieta, K., Brutovská, E. and Dušička, J. (2011c). Woody plants in evaluation of ecogenotoxicity and quality of the environment in the surroundings of industrial complexes. *Settlement - Park - Landscape* V, Nitra, SPU, 341-351.

Mičieta, K., Sámelová, A., Dušička, J. and Brutovská, E. (2013). Phytoindication of ecogenotoxicity in the proximity of Chemko Strážske. in: *Genetic Toxicology and Cancer Prevention*, Smolenice, 15-17.

Milacić, S., Petrović, D., Jovicić, D., Kovacević, R. and Simić J. (2004). Examination of the health status of populations from depleted-uranium-contaminated regions. *Environmental Research*, 95(1): 2-10.

Minouflet, M., Ayrault, S., Badot, P.-M., Cotelle, S. and Ferard, J.-F. (2005). Assessment of the genotoxicity of $^{137}Cs$ radiation using *Vicia*–micronucleus, *Tradescantia*–micronucleus and *Tradescantia*–stamen-hair mutation bioassays. *J. Environ. Radioactivity*, 81: 143-153.

Mišík, M. (2002). Individual and population plant response to the environmental pollution illustrated on the example of *Chelidonium majus* L. *Acta Fac. Rerum Nat. Univ. Comenianae*, Bot., 41: 9-14.

Mišík, M. (2007). Use of maize pollen grains assays in monitoring of environmental genotoxic impact. *Acta Bot. Univ. Comen.*, 43: 37-41.

Mišík, M., Mičieta, K. (2006). Impact of city pollution on micro-sporogenesis of *Chelidonium majus* L. plants. *Ekológia* (Bratislava), 25(1): 12-18.

Mišík, M., Solenská, M. (2004). Standardized phytoindicators in the detection of environmental mutagenesis under *in situ* conditions. In: Kontrišová, O., Ollerová, H., Váľka, J. (eds.), *Monitorovanie a hodnotenie stavu životného prostredia 5., časť A: Výsledky výskumu*, Zborník referátov z vedeckej konferencie, Technická univerzita, Zvolen, 151-159. (in Slovak) [*Monitoring and Evaluation of the Environment 5, Part A: Research Results*]

Mišík, M., Mičieta, K., Murín, G. and Solenská, M. (2003). Standardized test systems for the indication and monitoring of ecogenotoxicity *in situ*. In: Kočí, V., Maršálek, B., Halousková, O. (eds.), *Ecotoxicity Bioassays III,* Brno, Czech Republic, 154-157. (in Slovak)

Mišík, M., Solenská, M., Mičieta, K., Mišíková, K. and Knasmüller, S. (2006). *In situ* monitoring of clastogenicity of ambient air in Bratislava, Slovakia using the *Tradescantia* micronucleus assay and pollen abortion assays. *Mutat. Res.*, 605(1): 1-6.

Mišík, M., Mičieta, K., Solenská, M., Mišíková, K., Pisarčíková, H. and Knasmüller, S. (2007). *In situ* biomonitoring of the genotoxic effects of mixed industrial emissions using the *Tradescantia* micronucleus and pollen abortion tests with life plants: Demonstration of the efficacy of emission controls in an eastern European city. *Environ. Pollut.*, 145: 459-466.

Mišík, M., Ma, T.-H., Nersesyan, A., Monarca, S., Kim, J. K. and Knasmüller, S. (2011). Micronucleus assays with *Tradescantia* pollen tetrads: An update (Review). *Mutagenesis,* 26(1): 215-221.

Mišík, M., Kundi, M., Pichler, C., Filipic, M., Rainer, B., Mišíková, K., Nersesyan, A. and Knasmüller, S. (2016). Impact of common cytostatic drugs on pollen fertility in higher plants. *Environ. Sci. Pollut. Res.,* 23(15): 14730-14738.

Møller, A. P. (1998). Developmental instability of plants and radiation from *Chernobyl. Oikos,* 81: 444-448.

Mulcahy, D. L. (1981). Pollen tetrads as indicators of environmental mutagenesis. *Environ. Health Persp.*, 37: 91-94.

Mulcahy, D. L., Sari-Gora, M. and Mulcahy, G. B. (1996). Pollen selection – past, present and future. *Sex. Plant Reprod.*, 9: 353-356.

Mulder, C., Aldenberg, T., De Zwart, D., van Wijnen, H. and Breure, A. M. (2005). Evaluating the impact of pollution on plant-*Lepidoptera* relationships. *Environmetrics,* 16(4): 357-373.

Muratova, E. N., Sedel'nikova, T. S. (2004). Genomnye i chromosomnye mutacii u sosny obyknovennoj (*Pinus sylvestris* L.) v ekstremal'nych uslovijach proizrastenijach. [Genome and chromosome mutation in Scots pine (*Pinus sylvestris* L.) v extreme conditions of growth.] *Chvojnye borealnye zony,* 2: 128-139. (in Russian).

Murín, A. (1984). Simultaneous test of phytotoxic and mutagenic effects of polluted waters and herbicidal chemicals. *Biológia (Bratislava),* 39: 15-24.

Murín, A. (1987). Flowers as indicators of mutagenicity and phytotoxicity of polluted environment. *Biológia (Bratislava),* 42: 447-456.

Murín, A. (1995). Basic criteria for selection of plant bioindicators from the regional flora for monitoring of an environmental pollution. *Biológia (Bratislava),* 50: 37-40.

Murín A., Koleková A., Váchová M. and Regula Š. (1980). A simple guidance method for determining the phytotoxicity of chemically contaminated wastewater using model organisms of higher plants. *Biológia* (Bratislava), 35: 937-943. (in Slovak)

Murín, G. (1988a). Damage in the Structure and Reproduction of Chromosomes in Aging *Vicia faba* L. Seeds. *Acta F. R. N. Uni. Comen.* – Botanica, 35: 81-95.

Murín, G. (1988b). Individual Variability of Aging and Its Manifestations in *Vicia faba* L. Seeds. *Acta F. R. N. Univ. Comen.* – Botanica, 36: 87-95.

Murín, G. (1993). Danube water quality monitoring as part of the global biomonitoring network. In: Supeková. E., (ed.), *Dunaj-tepna Európy*, Bratislava, 108-115. (in Slovak) [*The Danube-Arteries of Europe*]

Murín, G. (1999). Biomonitoring of environmental pollution by indicative plant tests and sustainable development. In: Vilinovič, K., Chrenko, M. (eds.), *Trvalo udržateľné Slovensko?!*, 1. národná konferencia, Bratislava, 107-110. (in Slovak) [*Sustainable Slovakia?!*]

Murín, G., Malallah, G. (1994). Biomonitoring účinkov znečisteného životného prostredia po 100-dňovej vojne v Kuvajte. *Životné prostredie*, Bratislava, 1: 30-31. (in Slovak). [Biomonitoring of the effects of the polluted environment after a 100-day war in Kuwait. *Environment*]

Murín, G., Mičieta, K. (1993). New possibilities of plant tests of genotoxicity in international cooperation. *Environmental Liaison Centre*, Ecopoint, 3: 18-19.

Murín, G., Mičieta, K. (1996). Radiation impact on the selected wild plant species in the vicinity of nuclear power-plant Jaslovské Bohunice. *Folia Biol. (Praha),* 42(3): 122.

Murín, G., Mičieta, K. (1998). Pollen grains as an indicators of environmental stresses. in: Terek, O. I. et al. (eds.), *Proceedings of International conference "Plant ontogenesis in natural and transformed environments,"* Spolom Publ., Lviv, Ukraine, 271-272.

Murín, G., Mičieta, K. (2000a). Plant species for "*in situ*" microspore analysis of a polluted environment. *Environmental pollution and plant response.* CRC Press/Lewis Publishers, New York, 317-328.

Murín, G., Mičieta, K. (2000b). Wild plant species in bioindication of polluted environment. in: Yunus, M., Singh, N., De Kok, L. J., (eds.),

*Pollution stress: indication, mitigation and eco-conservation*, Kluwer Academic Publishers, 285-298.

Murín, G., Mičieta, K. (2001). The storage effect: An universal method of enhancing DNA's repair system. *Biologia,* 56/Suppl.10, 88pp.

Murín, G., Mičieta, K. (2008). Response of native flora to induceable genotoxic damage from increased radioactivity around NPP Jaslovské Bohunice, Slovakia. *International Symposium on Induced Mutation in Plants*, IAEA, Wienna, 56.

Murín, G., Mičieta, K. (2009a). Adaptive response of the wild plant species to the long-term pollution of the environment. *Ecotoxicology in the real world*, Kraków, Jagiellonian University, 81.

Murín, G., Mičieta, K. (2009b). Response of Plant Species to the Radioactive Contaminated Soil around Nuclear Power Plant Jaslovské Bohunice, Slovakia. In: Robert V. Steinberg, (Ed.). *Contaminated Soils: Environmental Impact, Disposal and Treatment*. New York: Nova Science Publishers, 433-448.

Murín, G., Mičieta, K. (2017a). Radioactively stimulated mutagenesis for a higher crop yield – past and present examples from Slovakia. *Acta Botanica Universitatis Comenianae,* Vol., 51: 59-62.

Murín, G., Mičieta, K. (2017b). The Rejuvenation by means of the Storage Effect: An Universal Method. *LAP LAMBERT* Academic Publishing, 168 pp.

Murín, G., Mičieta K., Knasmüller, S. and Steinkellner, J. (1996). Bioindication of radioactive-contaminated sites near nuclear power plant Jaslovské Bohunice by means of wild plant species. *Život. prostr.*, 3: 140-142.

Murín, G., Mičieta, K., Kunová, K. (1999). Simultaneous indication of phytotoxicity and genotoxicity of industrial wastes. in: Romančík, V. (ed.), *Industrial Toxicology* 1999, Bratislava, 81-84.

Murín, G., Murín, M. and Mičieta, K. (2004). Physical Indication of Aquatic Ecosystems. In: Kočí, V., Maršálek, B., Tlustá, P. (eds.), *Ecotoxicity Bioassays IV*, Chrudim, Czech Republic, 68-74. (in Slovak)

Murín, G., Mičieta, K., Chrenová, J. and Kovařík, P. (2007). Standardized plant tests for genotoxic phyto-indication of radioactive-contaminated soil from sites around NPP Jaslovské Bohunice. *Acta Fac. Ecologiae*, 16(1): 27-31.

Múdry, P., Gajdošová, A. (2010). Effect of radiomutagen on polymorphism of chosen enzymes in two amaranth species. In: *Nové poznatky z genetiky a šľachtenia poľnohospodárskych rastlín*. Zborník zo 17. vedeckej konferencie, Piešťany: VÚRV, 142–143. [*New knowledge of genetics and breeding of agricultural plants*.]

Müller, M., Zelling, G., Tausz, M., De Kok, L. J. and Grill, D. (1997). Root responses of forest tress to environmental stresses. In: Responses of plant metabolism to air pollution and global change. In: *IV. Int. Symp., Egmond.*, 47.

Müller, M., Tausz, M., Wonisch, A., Guttenberger, H. and Grill, D. (2000). Evaluation of chromosomal aberrations in the root tip meristems of spruce trees for the assessment of environmental influences on forest tree species. In: Guttenberg, H., Borzan, Z., Schlarbaum, S.E. & Hartman, T. P. V. (eds.). *Cytogenetic Studies of Forest trees and Shrubs – Review, Present Status, and Outlook on the Future. Special Issue of the Forest genetics.* Arbora Publishers, Zvolen, Slovakia, 121-127.

Natarajan, A. T., Obe, G. (1980). Screening of human populations for mutations induced by environmental pollutants: Use of human lymphocyte system. *Toxicol. Environ. Saf.,* 4: 468-81.

Nebish, A. A. (2004). Estimates of genotoxic effect by the analysis of male gametophyte generation of vines. *Proceedings of the International Conference URM.* Yerevan, Armenian National Academy of Science, 55.

Nicol, J. A. C., Donahue, W. H., Wang, R. T. and Winters, K. (1977). Chemical composition and effects of water extracts of petroleum on eggs of the sand dollar, *Mellita quinquesperforata. Mar. Biol.*, 40: 309-316.

OECD (1984). Guideline for testing of chemicals 208. *Terrestrial plants,* Growth test.

OECD (2000). *Guideline for testing of chemicals. Proposals for updating Guideline 208. Terrestrial (non-target) plant test: 208A: Seedling emergence and seedling growth test. 208 B: Vegetative vigour test.* Draft document.

Ostrolucká, M. G., Fleischer, P. (2003). Content of heavy metals in pollen and seeds of Norway spruce (*Picea abies* (L.) Karst.) in central part of the High Tatra mts. *Biológia*, Bratislava, 22(1): 65–73.

Ostrolucká, M. G., Bolvanský, M. and Tokár, F. (1997). Seed production and seed viability in black pine (*Pinus nigra* Arnold) at sites with different level of pollution. *Biológia*, Bratislava, 16(2): 129–136.

Oudalova, A., Geraskin, S., Vasiliev, D. and Dikarev, V. (2005). Cytogenetic variability in *Pinus sylvestris* L. populations experiencing anthropogenic influence. *Radioprotection, Suppl. 1,* 40: 223-228.

Öztürk, M., Aksoy, M. A. A. and Ahmad, M. S. A. (eds) (2015). *Phytoremediation for Green Energy*, Springer, 191pp.

Paradiž, J., Lovka, M. (1999). Pollen Grain Bioassay for Environmental Pollution Screening. *Phyton – Annales Rei Botanicae*, 39: 175-182.

Parantainen, A., Pulkkinen, P. (2002). Pollen viability of Scots pine (*Pinus sylvestris*) in different temperature conditions: high level of variation among and within latitudes. *Forest Ecology and Management*, 167: 149–160.

Pavlova, D. K. (2016). Effect of nickel on pollen germination and pollen tube length in *Arabis alpine* L. (*Brassicaceae*). *Australian Journal of Botany*, 66-69.

Pavlova, D., De La Fuente, V., Sánchez-Mata, D. and Rufo, L. (2014). Pollen morphology and localization of Ni in some Ni-hyper-accumulator taxa of *Alyssum* L. (*Brassicaceae*) *Plant Biosystems*, 12: 1-11.

Pavlova D., Laporte, F., Ananiev, E. D. and Herzog, M. (2016). Pollen Morphological Studies on *Arabis alpina* L. (*Brassicaceae*) Populations from the Alps and the Rila Mountains. *Genetics and Plant Physiology*, 6(1-2): 27–42.

Peška, J. and 34 authors (1980). Využití nukleárních metod a ionizujícího záření v genetice, šlechtění a fyziologii rostlin. In: J. Peška (Chief Ed.).

*Proceedings of a conference*, Brno: Vysoká škola zemědělská, 194 pp. [Use of nuclear methods and ionizing radiation in genetics, breeding and plant physiology]

Pfahler, P. L. (1975). Factors affecting male transmission in maize (*Zea mays* L.). In: D. L. Mulcahy (Ed.) *Gamete competition in plants and animals*, Elsevier Press, Amsterdam, 115-124.

Pichler, C., Filipič, M., Kundi, M., Rainer, B., Knasmüller, S. and Mišík, M. (2014). Assessment of genotoxicity and acute toxic effect of the imatinib mesylate in plant bioassays. *Chemosphere,* 115(1): 54-58.

Pipíška, M., Lesný, J., Horník, M. and Augustín, J. (2004). Plant uptake of radiocesium from contaminated soil. *Nukleonika*, 49(1): 9-11.

Pisarčíková, H., Mičieta, K., Šípková, Ž, Solenská, M. and Murín, G. (2006). Indication of ecogenotoxicity deterioration of significant woody species in some Slovak regions. in: Kočárek, Plášek, V., Malachová, K. (eds.), *Environmental changes and biological assessment III.* Scripta Facultatis Rerum Naturalium Univesitatis Ostraviensis, 163: 339-346.

Plewa, M. J. (1982). Specific-locus mutation assays in *Zea mays*. *Mutat. Res*., 99: 317-337.

Pogányová, A., Kerekeš, E. and Mičieta, K. (2017). The ecogenotoxic plant biomonitoring of a long-term polluted area in central Slovakia. *Environmental Sci. Pollution Res.,* 24: 27376-27383.

Posthumus, A. C. (1983). Higher plants as indicators and accumulators of gaseous air pollution. *Environ. Monit. Assess.,* 3: 263-272.

Posthumus, A. C., Tonnejick, A. E. G., (1982). Monitoring of effect of photo-oxidants on plants. in: Steubing, L., Jäger, H. J. (eds.), *Monitoring of air pollutants by plants. Methods and problems*, Dr. W. Junk Publishers, The Hague, 115-119.

Pozolotina, V.N. (2001). Analysis of local dandelion (*Taraxacum officinales*. l.) cenopopulations from radioactively contaminated zones. *Russ. J. Ecol*., 32(2): 102–109.

Prus-Glowacki, W., Godzik, S. (1991). Changes induced by zinc smelter pollution in the genetic structure of pine (*Pinus sylvestris* L.) seedling populations. *Silvae genetica*, 40: 184-188.

Prus-Glowacki, W., Novak-Bzowy, R. (1992). Genetic structure of a naturally regenerating Scots pine population tolerant for high pollution near a zinc smelter. *Water Air Soil Poll.*, 62(3-4): 249-259.

Pukacki, P. M., Chałupka, W. (2003). Environmental pollution changes in membrane lipids, antioxidants and vitality of scots pine (*Pinus sylvestris* L.) pollen. *Acta Societatis Botanicorum Poloniae*, 72(2): 99–104.

Qiang, L., Liu, Fu-P. (2003). A review of phytoremediation to chemical air pollution and plant species selection for greening. *Subtropical Plant Science*, 32(3): 20-29.

Rai, P. K. (2013). Environmental magnetic studies of particulates with special reference to biomagnetic monitoring using roadside plant leaves (Review). *Atmospheric Environment*, 72: 113-129.

Rashad, S. M., Hammad, F. H. (2000). Nuclear power and the environment: comparative assessment of environmental and health impacts of electricity – generating systems. *Applied Energy*, 65: 211-229.

Rea, M. A., Rodríguez-Muňoz, Ma. E., Rico-Rodríguez, M. A., Anaya-Alonso, A. L. and Grantz, D. A. (2006). Brickmaking in Agricultural Communities in Mexico: Distribution, Fuels, Inventory, Emissions, and Effects on Animals and Plants. *Workshop on Agricultural Air Quality*, Washington, D.C., 1023-1032.

Readman, J. W., Fowler, S. W., Villeneuve, J. P., Cattini, C., Oregioni, B. and Mee, L. D. (1992). Oil and combustion-product contamination of the Gulf marine environment following the war. *Nature,* 358: 662-664.

Regis, E. G., (1999). *Pollen grain abortion as indicator of mercury pollution near gold mining sites in Camarines Norte*. A doctoral dissertation under the PhD Environmental Science Program of the College of Science, University of the Philippines, Diliman, Quezon City. Philippines, 155pp.

Regis, E. G., Lagunzad, D. A. (2002). Comparison of pollen abortiveness in four weed species treated with mercuric chloride. *Science Diliman*, 14(1): 21-27.

Regis, E. G., Medrano, I., Ma. Aurora Azurin-Conde and Melanio Regis (2001). Some Impacts of Mining on the Island Ecosystem of Rapu-Rapu, Province of Albay. *Gibon*, 1(1): 43-77.

Regis, E. G., Barrameda, L. and Montañez, M. A. (2006). New methods for detecting heavy metals in mining sites developed. Ateneo de Naga University, *New method*, 51-54.

Renzoni, A. (1975). Toxicity of three oils to bivalve gametes and larvae. *Mar. Pollut. Bull.*, 6: 125-128.

Renzoni, G. C., Viegi, L. (1991). In vitro sensitivity of *Pinus pinaster* and *P. pinea* pollen grains to different pH values. *Ann. Bot. Fennici*, 28: 135-142.

Renzoni, G. C., Viegi, L., Stefani, A. and Onnis, A. (1990). Different in vitro germination responses in *Pinus pinea* pollen from two localities with different levels of pollution. *Ann. Bot. Fennici*, 27: 85-90.

Reponen, T., Grinshpun, S. A., Trakumas, S., Martuzevicius, D., Wang, Z. M., LeMasters, G., Lockey, J. E. and Biswas, P. (2003). Concentration gradient patterns of aerosol particles near interstate highways in the Greater Cincinnati airshed. *Journal of Environmental Monitoring*, 5: 557–562.

Rizk, R. M., Soliman, M. I., Baka, Z. A. and Al-Saoud, A. M. A. (2015). Assessment of Negative Environmental Impacts of Drainage and Irrigation Canals on *Ipomoea carnea* Jacq. in *Egypt. International Journal of Plant Breeding and Genetics*, 9(3): 143-162.

Samet, J. M., DeMarini, D. M. and Malling, H. V. (2004). Do Airborne Particles Induce Heritable Mutations? *Science,* 304: 971-972.

Sámelová, A., Mičieta, K., Brutovská, E. and Dienes, K. (2011). Physical indication of ecogenotoxicity of old environmental loads on the example of gudron dumps and gas generator storage in Kalinkovo village. *Acta Environmentalica Universitatis Comenianae* (Bratislava), 19(1): 305-308. (in Slovak)

Sandhu, S. S., De Serres, F. J., Gopalan, H. N. B., Grant, W. F., Veleminský, J. and Becking, G. C. (1991a). Status report of the International Programme on Chemical Safety's Collaborative study on plant test systems. *Mutat. Res.,* 257: 19-25.

Sandhu, S. S., Gill, B. S., Casto, B. C. and Rice, J. W. (1991b). Application of *Tradescantia* micronucleus assay for *in situ* evaluation of potential genetic hazards from exposure to chemicals at a wood-preserving site. *Waste Hazard*, 8: 257-262.

Sandhu, S. S., De Serres, F. J., Gopalan, H. N. B., Grant W. R., Velemínský J. and Becking G. C. (1994a). Environmental monitoring for genotoxicity with plant systems: An introduction and study design. *Mutat. Res.*, 310(2): 169-173.

Sandhu, S. S., De Serres, F. J., Gopalan, H. N. B., Grant, W. R., Svendsgaard, D., Velemínský, J. and Becking, G. C. (1994b). Environmental monitoring for genotoxicity with plant systems: Results and recommendations. *Mutat. Res.*, 310(2): 257-263.

Samecka-Cymerman, A., Kolon, K., Mróz, L. and Kempers, A. J. (2012). Bioindicative comparison of the fern *Athyrium distentifolium* for trace pollution in the Sudety and Tatra mountains of Poland. *Environ. Monit. Assess.*, 184(10): 6357-6365.

Seibert, H., Gulden, M., Kolossa, M. and Schepers, G. (1992). Evaluation of the relevance of selected in vitro toxicity test systems for acute systemic toxicity. *ATLA,* 20: 240-245.

Schlarbaum, S. E. (2000). Cytogenetic studies of forest trees: Looking to the past to meet challenges in the future. In: Guttenberg, H., Borzan, Z., Schlarbaum, S. E. & Hartman, T. P. V. (eds.). *Cytogenetic Studies of Forest trees and Shrubs- Review, Present Status, and Outlook on the Future.* Special Issue of the Forest genetics. Arbora Publishers, Zvolen, Slovakia. 9-19.

Schnoor, J. L. (1997). Phytoremediation. Technol. Eval. Report TE-98-01. Pittsburg, *Ground-Water Remed. Tech. Anal. Center*, 37 pp.

Schnoor, J. L. (2000). Degradation by plants – Phytoremediation. In: Klein, J. (Ed.), *Biotechnology*. Wiley-VCH,Weinheim, 11b: 372-384.

Sharma, C. B. S. R., Raju, D. S. S. (1983). *In situ* cytogenetic monitoring of vegetation around Hindustan polymers factory. Vishakhapatnam, India. *Environ. Mutagens*, 5: 409-417.

Sheoran, I. S., Saini, M. S. (1996). Drought-induced male sterility in rice: Changes in carbohydrate levels and enzyme activities associated with the inhibition of starch accumulation in pollen. *Sex Plant Reprod.*, 9: 161-169.

Shevchenko, V. A. (1989). Some Aspects of the Genetic Consequences of the Chernobyl Disaster. Transactions of the *10th International conference on Structural Mechanics in reactor technology*, Vol. D., Los Angeles, 245-249.

Sheppard, S. C. (2003). An index of radioactivity, what has been important? *J. Environ. Radioactivity*, 68(1): 1-10.

Shull, J. K., Meazel, M. Y. (1977). A study of the reliability of synchrony in the development of pollen mother cells of *Lillium longiflorum*. *Am. J. Bot.*, 64: 670-679.

Simons, M. (2001). Radiation from Balkan Bombing Alarms Europe. *NY Times*, Jan. 7., 23-27.

Sheppard, S. C., Guthrie, J. E., Thibault, D. H. (1992). Germination of Seeds from an Irradiated Forest: Implications for Waste Disposal. *Ecotoxic. and Environ. Safety*, 23: 320-327.

Slávik, O., Morávek, J. (2000). Measuring techniques for the characterization of $^{137}$Cs contaminated river banks, Slovakia. in: *Site characterization techniques used in environmental restoration activities,* IAEA-TECDOC, 1148: 153-171.

Słomka, A., Jedrzejczyk-Korycińska, M., Rostański, A., Karcz, J., Kawalec, P., Kuta, E. (2012). Heavy metals in soil affect reproductive processes more than morphological characters in *Viola tricolor*. *Environmental and Experimental Botany*. 75: 204-211.

Słomka, A., Bohdanowicz, J., Poznańska, E., Kwiatkowska, M., Pilarska, M., Struebig, M., Shuka, L. and Kuta, E. (2014). Usefulness and limitations of pollen characters in environmental studies based on *Viola* L. species (sect. Melanium Ging.). *Modern. Phytomorphol.*, 5: 31-34.

Solenská, M. (2005). Microsporogenesis of higher plants in monitoring and detection of genotoxical effects. *Acta Bot. Univ. Comen.*, 42: 45-49.

Solenská, M., Mičieta, K. and Mišík, M. (2003). Phytoindication of ecogenotoxicity in industrial area in Žilina. *Industrial Toxicology*, Bratislava, 88-94.

Solenská, M., Mišík, M. and Mičieta, K. (2005a). *Tradescantia* mikrojadrový test v *in situ* monitoringu genotoxicity ovzdušia priemyselne zaťažených oblastí. in: *Ovzduší*, 158-161. (in Slovak).

Solenská, M., Mišík, M. and Mičieta, K. (2005b). Evaluation of soil genotoxicity from industrial area using *Tradescantia* micronuclei test and *Vicia faba* bioanalysis. in: *Študentská vedecká konferencia* 2005, 1. zväzok – biologická sekcia, Bratislava, Kartprint, 118. [*Student Scientific Conference*]

Solenská, M., Mičieta, K. and Mišík, M. (2005c). Microsporogenesis of wild plants species in monitoring of genotoxicity in pollutant impacted city area – four years study. in: *Industrial Toxicology*, Bratislava, 73-77.

Solenská, M., Mičieta, K. and Mišík, M. (2006). Plant bioassays for an *in situ* monitoring of air near an industrial area and a municipal solid waste - Žilina (Slovakia). *Environ. Monit. Assess.*, 115(1-3): 499-508.

Solenská, M., Mičieta, K., Mišík, M. and Murín, M. (2007). Ecotoxicity and genotoxicity in wastewater samples discharged to small Danube and Danube river in Bratislava. *Industrial Toxicology*,Bratislava, Slovak Republic, 49-59.

Somashekarappa, H. M., Narayana, Y., Radhakrishna, A. P., Karunakara, N., Balakrishna, K. M. and Siddappa, K. (1996). Bioindicators in the Tropical Forest of Kaiga Environment. *J. Environ. Radioactivity*, 31(2): 189-198.

Somers, C. M., McCarry, B. E., Malek, F. and Quinn, J. S. (2004). Reduction of Particulate Air Pollution Lowers the Risk of Heritable Mutations in Mice. *Science*, 304: 1008-1010.

Sorkhoh, N., Al-Hasan, R., Radwan, S. and Höpner, Th. (1992). Self-cleaning of the Gulf. *Nature*, 359: 109-109.

Sorochinskii, B. V. (1998). The characteristics of the protein composition of the anomalous needles of spruce (*Picea abies*) and pine (*Pinus sylvestris*) from the 10-kilometer zone of the Chernobyl Atomic Electric Power Station. *Tsitol Genet.*, 32(5): 35-40.

Sorochinskii, B. V. (2003). Molecular-biological nature of morphological abnormalities induced by chronic irradiation in coniferous plants from the Chernobyl nuclear power plant exclusion zone: emphasis on a possible role of the cytoskeleton. *Tsitol. Genet.*, 37(2): 49-55.

Sorochinskii, B. V., Ruchko, M. V. and Prokhnevsky, A. I. (1998). Irradiation-induced abnormal morphogenesis of coniferous plants at the Chernobyl exclusion zone and it's possible mechanisms. in: Terek, O. I., Berko, J. V., Volph, S. O., Vorobec, N. M., Demkyv, O. T., Krinickij, G. T., Romanjuk, N. D., Cvilinjuk, O. M. (Eds.), *Proceedings ofInternational conference on plant ontogenesis in natural andtransformed environments*, Lviv, Ukraine, 286-287.

Sparling, D. W. (1995). Acidic deposition: A review of biological effects. In: Hoffman, D. J., Rattner, A. B., Burton, G. A. and Cairns, J. (eds.). *Handbook of Ecotoxicology*. Lewis Publishers, London-Tokyo, 301-329.

Speranza, A., Leopold, K., Maier, M., Taddei, A. R. and Scoccianti, V. (2010). Pd-nanoparticles cause increased toxicity to kiwifruit pollen compared to soluble Pd (II). *Environ. Pollut.,* 158: 873–882.

Sracci, R., Kogevinas, M., Bertazzi, P. A., Bueno de Mesquita, B. H., Coggon, D. and Green, L. M. (1991). Editorial, *The Lancet*, 338: 8774.

Sriussadaporn, C., Yamamoto, K., Fukushi, K. and Simazaki, D. (2003). Comparison of DNA damage detected by plant comet assay in roadside and non-roadside environments. *Mutation Research*, 541: 31-44.

Sulleiman M. K., Bhat N. R. (2003). Performance of ornamental plants in bioremediated soil. *Arid Land Res. Manag.* 17(2), 169-176.

Svetková, K., Mičieta, K. and Fargašová, A. (2005). Use of higher plants for genotoxicity assessment of waste water polluted by heavy metals. *Industrial Toxicology*, Bratislava, 338-342.

Šeliga, M. (2006). Ďalšie aplikácie ionizujúceho žiarenia. In: Miroslav Tibor Morovics (Chief Ed.). *Atómy na Slovensku*, Slovenská nukleárna spoločnosť, 135-146 (in Slovak) Other applications of ionizing radiation. [*Atoms in Slovakia*]

Šípková, Ž., Mičieta, K. (2005). Pollen aboritivity in populations of *Polemonium caeruleum* L. in connection with genotoxic impact of traffic air pollution. *Acta Bot. Univ. Comen.*, 42: 49-55.

Štamberg, K, Vopálka, D., Beneš, P. and Slávik, O. (1997). A simplified approach to modeling underground migration of radionuclides from contaminated river sediments. *J. Contam. Hydrol.*, 26(1-4): 301-307.

Štrba, P., Benková, J. (2011). Diverzita cievnatých rastlín na banských haldách. *Vliv abiotických a biotických stresorů na vlastnosti rostlin*, Praha–Ruzyně, VÚRV, 154. (in Slovak) Diversity of vascular plants at mining heaps. [Diversity of vascular plants at mining heaps. *Influence of abiotic and biotic stressors on plant properties*]

Štubňa, M., Pekár, A., Morávek, J. and Špirko, M. (2002). Decommissioning Project of Bohunice A1. *Conference Waste Management 2002 Symposium*, Tucson, Arizona (US), 1-12.

Theodorsson-Norheim, E. (1986). Kruskal-Wallis test – Basic computer program to perform nonparametric one-way analysis of variance and multiple comparisons on ranks of several independent. *Comp. Meth. Prog. Biomed.*, 23: 57-62.

Tomkins, D. J., Grant W. F. (1976). Monitoring natural vegetation for herbicide-induces chromosomal aberrations. *Mutat. Res.,* 36: 73-84.

Tretyakova, I. N., Bazhina, E. V. (2000). Structure of crown as well as pollen and seed viability of *Abies sibirica* (Ledeb.) in disturbed forest ecosystems of the Khamar-Daban mts near Baikal lake. *Ekológia*, Bratislava, 19(3): 280–294.

Tretyakova, I. N., Noskova, N. E. (2004). *Scotch pine* pollen under conditions of environmental stress. *Russ. J. Ecol.*, 35(1): 20-26.

Tulik, M., Rusin, A. (2005). Microfibril angle in wood of Scots pine trees (*Pinus sylvestris*) after irradiation from the Chernobyl nuclear reactor accident. *Environ. Pollution*, 134: 195-199.

Udosen, E. D., Benson, N. U., Essien, J. P. and Ebong, G. A. (2006). Relation between aqua-regia extractable heavy metals in soil and *Manihot utilissima* within a municipal dumpsite. *Int. J. Soil Sci.*, 1(1): 27-32.

Uhríková, A., Mičieta, K. (1995). *In situ* bioindication of genotoxic effect using the species of native flora in the vicinity of nickel plant. *Biológia* (Bratislava), 50: 65-68.

UNITAR (2013). https://www.unitar.org/bomb-survivor-tree-spread-peace-message.

UNSCEAR (2008). http://www.unscear.org/unscear/en/chernobyl.html.

U.S. EPA. (1985). TSCA SS797.2750, SS7972800. Seed germination/root elongation. *Early seedling growth toxicity test.*

U.S. EPA. (1996). *OPPTS Ecological effect guideline*, Series 850.4000–850.4230.

Vačková, S. (2002).Variability of chromosomal aberration frequency of *Polygonatum multiflorum* and *P. latifolium* species under the different environmental conditions. *Acta Fac. Rerum Nat. Univ. Comenianae,Bot.,* 41: 36.

Valencia-Quintana, R., Gómez-Arroyo, S., Waliszewski, S. M., Sánchez-Alarcón, J., Gómez-Olivares, J. L., Flores-Márquez, A. R., Cortés-Eslava, J. and Villalobos-Pietrini, R. (2012). Evaluation of the genotoxic potential of dimethyl sulfoxide (DMSO) in meristematic cells of the root of *Vicia faba. Toxicol. Environ. Health Sci.*, 4(3): 154-160.

Van't Hof, J., Schairer, L. A. (1982). *Tradescantia* assay system for gaseous mutagens. *Mutat. Res.,* 99: 303-315.

Vig, B. K. (1982). Soybean [*Glycine max* L. (Merill)] as a short-term assay for study of environmental mutagens. *Mutat. Res.*, 99: 339-349.

Volkova, P. Yu., Geras'kin, S. A., Horemans, N., Makarenko, E. S., Saenen, E., Nauts, R., Bondarenko, V. S., Duarte, G.T., Jacobs, G., Voorspoels, S. andKudin, M. (2018). Chronic radiation exposure as an ecological factor: hypermethylation and genetic differentiation in irradiated Scots pine populations. *Environmental Pollution*, 232: 105-112.

Vyshenska, I. G., Som, O. V. (2001). Bioindication of Radioactive Contamination of Teritories by the Pollen Sterility Analysis. *Naukovi Zapiski*, 19: 74-76.

Wolters, J. H. B., Martens, M. J. M. (1987). Effect of air pollution on pollen. *Bot. Rev.*, 53: 372-414.

Závodský, D. (2007). Znečistenia ovzdušia Bratislavy v rokoch 1965 – 2005. In: Střelcová, K., Škvarenina, J. Blaženec, M. (Eds.). *Bio-*

*climatology and Natural Hazards,* International Scientific Conference, Poľana nad Detvou, Slovakia, 12pp. (in Slovak) Air Pollution of Bratislava in the years 1965 – 2005.

Zaka, R., Vandecasteele, C. M. and Misset, M. T. (2002). Effects of low chronic doses of ionizing radiation on antioxidant enzymes and $C_6$ PHD activities in *Stipa capillata* (Poaceae). *J. Exp. Bot.*, 53: 1979-1987.

Zelena, L., Sorochinsky, B., von Arnold, S., van Zyl, L. and Clapham, D. H. (2005). Indications of limited altered gene expression in *Pinus sylvestris* trees from the Chernobyl region. *Journal of Environmental Radioactivity*, 84: 363-373.

Zeleňáková, E., Holub, Z. (1985). Ecological problems in the imission area of nickel works. *Agriculture (Nitra),* 31: 1003-1008.

# About the Authors


## *Karol Mičieta, PhD and Gustáv Murín*[*]*, PhD*

Department of Botany, Faculty of Natural Sciences,
Comenius University, Bratislava, Slovak Republic


***Karol Mičieta***, Prof. RNDr, PhD, born November, 19th, 1952, in Žilina, Slovakia, is educated at Comenius University (UK), Bratislava, Slovakia, as systematic botanist. Doctor honoris causa is long time head of Department of Botany at Comenius University and currently a rector of this University. Member of several Scientific Councils and commissions in Slovakia and Czech Republic. Holder of the Silver medal of Faculty of Natural Sciences, UK; Golden medal of the Slovak Agricultural University, Nitra; Galen Memorial Medal of Faculty of Pharmacy, UK, and Silver commemorative medal, UK. Pedagogic achievement: 7 PhD students and 55 diplomants. Publications: monographs 3x, chapters in foreign monographs 5x and journal publications: 475x. Citation response: more than 400x SCI.


[*] Corresponding Author Email: gustavmurin@yahoo.com

***Gustáv Murín***, RNDr, PhD, born April, 9th, 1959, in Bratislava, Slovakia, is educated at Comenius University, Bratislava, Slovakia, as cytologist. Fellowships: at the Inst. of Exp. Bot., Prague; at the Biophysical Inst., Brno; at Genetic Tox., Swansea, UK; at the Inst. für Pflanzengenetik und Kulturpflanzenforschung in Gatersleben, Germany; at Dep. Biologia Celular, Fac. de Biologia, Univ. de Sevilla, Spain; at Inst. für Tumorbiologie-Krebsforschung Der Univ. Wien, Austria; at the Dep. of Mol. Biol., Univ. of Zagreb, Croatia, and at the Dep. of Plant Sci., Clare College, Cambridge, UK. Stipend of Hoallender Comm. of the Envir. Mutagen Society, Japan. Publications: monographs 2x, chapters in foreign monographs 5x and journal publications: 111. Citation response: 121.

# INDEX

## A



## B

## C



## D

## G



## H



## I



## J

## K



## L



## M



## N

## O



## P

## Q



## R



## S